Physics, Structure, and Reality

JILL NORTH

OXFORD
UNIVERSITY PRESS

OXFORD
UNIVERSITY PRESS

Great Clarendon Street, Oxford, OX2 6DP,
United Kingdom

Oxford University Press is a department of the University of Oxford.
It furthers the University's objective of excellence in research, scholarship,
and education by publishing worldwide. Oxford is a registered trade mark of
Oxford University Press in the UK and in certain other countries

First Edition published in 2021

Impression: 1

Published in the United States of America by Oxford University Press
198 Madison Avenue, New York, NY 10016, United States of America

British Library Cataloguing in Publication Data
Data available

Library of Congress Control Number: 2021935065

ISBN 978–0–19–289410–6

DOI: 10.1093/oso/9780192894106.001.0001

Printed and bound in the UK by
TJ Books Limited

for Mom and Dad

Preface

Thanks to friends and colleagues for invaluable feedback at various stages: Valia Allori, Thomas Barrett, Ori Belkind, Gordon Belot, Karen Bennett, Jim Binkoski, Carolyn Brighouse, Alexander Ehmann, Nina Emery, Ned Hall, Hans Halvorson, Thomas Hofweber, Chris Meacham, Alyssa Ney, Laura Ruetsche, Juha Saatsi, Ted Sider, Jim Weatherall, Isaac Wilhelm, and Mark Wilson. (Many apologies to anyone I may be forgetting.) Thanks to the National Science Foundation for funding that supported the research for Chapter 5 and more. Thanks to Oxford University Press for permission to reprint some material from "A New Approach to the Relational-Substantival Debate," which appeared in *Oxford Studies in Metaphysics* Volume 11. The epigraph to Chapter 3 is taken from *Fact, Fiction, and Forecast* by Nelson Goodman, Cambridge, Mass.: Harvard University Press, Copyright © 1979, 1983 by Nelson Goodman, reproduced by permission of the publisher. Many thanks to those at or affiliated with Oxford University Press: Henry Clarke, S. Kabilan, Vaishnavi Anantha Subramanyam, Matthew Williams, copy editor Kim Allen, and especially Peter Momtchiloff.

A huge thanks as well, more indirectly, to my teachers over the years: I am very grateful.

Contents

1
Introduction

> Many of the scientific treatises of today are formulated in a half-mystical language, as though to impress the reader with the uncomfortable feeling that he is in the permanent presence of a superman. The present book is conceived in a humble spirit and is written for humble people.
>
> Cornelius Lanczos (1970, vii–viii)

Our best physical theories are formulated in abstract mathematical terms. This creates difficulties for the interpretive project of figuring out what these theories are saying about the world. Simply put, it is not obvious what a mathematical formalism says about the physical world. Of course, any mathematical formulation we devise will be based on pieces of experimental evidence and manifestly observable features of the world; but those things will not pin down the full nature of the world according to a theory. There is no conclusive rule or algorithm that takes us from a mathematical formalism to the nature of the physical world, and yet we do seem able to draw reasonable conclusions about the world on the basis of these theories.

How do we do this? How do we figure out the nature of the world from a mathematically formulated physical theory? And what if a theory can be formulated mathematically in different ways, as is typically the case—what do we infer about the world then?

This book is about this interpretive project. It is about the relationship between the mathematical structures in which our physical theories are formulated, and the nature of the physical world(s) these theories describe. I will be suggesting that there is a certain notion of structure that is familiar (if often inexplicit) in physics and mathematics, and that paying attention to structure in this sense, both in the mathematical formalism and in the physical world, is important to figuring out what physics, especially fundamental physics, is saying about the world.

There has been lots of discussion in philosophy recently, in philosophy of science and philosophy of physics in particular, centering on various notions of structure. Let me say a little about this by way of contrast with my own ideas.

The notion that is most familiar in philosophy of science comes from the literature on structural realism, stemming from the discussion of John Worrall (1989) (who finds similar ideas in Poincaré). Worrall uses a particular conception of structure to respond to the so-called pessimistic meta-induction against scientific

Physics, Structure, and Reality. Jill North, Oxford University Press (2021). © Jill North.
DOI: 10.1093/oso/9780192894106.003.0001

realism. This is an argument claiming to show that we have no reason to believe that our scientific theories are getting at the truth about the world, as the realist maintains. As a matter of historical fact, many scientific theories that had been successful turned out to be false. Inductive reasoning from these past failures then suggests that we should not believe in the truth of our current theories either: it is likely that these theories, too, will eventually be shown to be false. Since the past is littered with falsified yet successful scientific theories, it seems as though we have no reason to believe that our current theories, successful as they are, are even getting any closer to the truth.

The view that has come to be called "structural realism" aims to acknowledge this fact about past scientific theories, while at the same time allowing the realist to accept the "no miracles" argument *for* scientific realism—that the success of our scientific theories would be a miracle if they weren't true, or at least approximately true. The structural realist says that even though many past scientific theories have been abandoned as false, there is a *structure* to these theories that remains in place throughout the process of theory change, and about which we can be realists. As Worrall says of the shift from Fresnel's to Maxwell's theory of light, for example, the former seeing light as mechanical vibrations through a solid elastic medium, the latter as waves in an electromagnetic field: "There was continuity or accumulation in the shift, but the continuity is one of *form* or *structure*, not of content" (1989, 117; original italics). Fresnel was wrong about the nature of light; yet his theory was empirically successful because it hit upon the correct structure, in the sense of the correct form of equations governing light's transmission.

According to the structural realist, our scientific theories are getting at the truth about the structure of the world, if not its intrinsic nature. Structure is the kernel of truth that successful scientific theories have in common, and about which we can safely be realists even in the face of the pessimistic meta-induction. One version of the view, known as epistemic structural realism, says that scientific theories give us knowledge about the structure of the world rather than its intrinsic nature. Another version, ontic structural realism, goes further to say that all there is to reality is structure; or that structure is primary or fundamental, and objects are secondary or nonfundamental. This view urges "a shift in one's ontology, away from objects, as traditionally conceived, and towards structures, typically conceived of in terms of relations," so that "inasmuch as objects exist at all, they derive their properties and individuality from the relational network in which they are embedded" (Rickles and French, 2006, 25; 4). The suggestion is that we must abandon a standard object-oriented metaphysics, as in James Ladyman and Don Ross's book *Every Thing Must Go* (2009), the title of which conveys the gist.[1]

[1] See Ladyman (2016) and McKenzie (2017) and references therein for discussion of all these ideas and variations on them.

Other uses of structure include a view in philosophy of physics called spacetime structural realism, which holds that spacetime in particular is nothing but a certain kind of relational structure. This view claims to be an alternative to the traditional relational-substantival dichotomy, allowing us to circumvent various difficulties that arise from taking spacetime points to be additional elements in the ontology: we are to believe in the existence of a relational spacetime structure without having to commit to the existence or fundamentality of spacetime points as objects. Other conceptions of structure in recent philosophy of physics are based explicitly on mathematical notions from model theory or category theory (as in Halvorson, 2019), which are then put to use in addressing various topics in philosophy of science, such as the nature of theoretical equivalence. Ted Sider (2011) argues that a particular notion of structure, distinct from the one prevalent in recent philosophy of science and physics, is fruitful throughout various areas of philosophy, helping to clarify what is at issue in many traditional philosophical puzzles and debates.

It may seem like the last thing we need is another treatise on this well-worn notion. I hope to convince you otherwise. It is true that the term "structure" has been bandied about, but it is used in very different ways by different people and in the contexts of different philosophical debates. I intend it in a particular way that is implicit in various aspects of our theorizing about physics, and is distinct from the other notions on tap. Not completely distinct, mind you: there is a reason for the common usage. But it is different enough. It is something that, taken seriously, yields progress on the questions posed in the second paragraph. That this notion is distinct will become clear through the following chapters. One significant difference is that my conception of structure does not have anything particularly to do with the existence or relative priority of objects or intrinsic properties as opposed to relations or relational structures. Nor is my chief concern to try to salvage realism in the face of the pessimistic meta-induction. (I do not in any case find that argument very compelling. It seems to me that we can accept the history of abandoned theories while still having good reason to believe that our current theories, with all the additional evidence we have for them, are getting closer to the truth, though I won't argue the point here.[2]) I will be advocating a realism about structure, but one that is quite different from what currently goes by the name. My overall aim is thus to further this "structural turn," while at the same time distancing my approach from other notions and uses of structure in recent literature.

For the rest of this introductory chapter, I want to sketch my general outlook and some motivating themes, to help situate the discussion within the philosophical

[2] Valia Allori has generously suggested that my view should contain the ingredients to respond to that argument. That may be, but I am not sure. Although my account may indirectly bolster the case for scientific realism, I am not sure that it deflects the pessimistic meta-induction in particular in a way that people bothered by the argument will be happy with. (For instance, I do not argue that there must be some common structure that is preserved through theory change.) I will leave it to the experts to say whether my ideas have anything novel to offer by way of defusing the pessimistic meta-induction.

landscape. Some of these themes will be argued for, others will form background assumptions. These themes will be controversial to varying extents. One thing I hope to eventually show, however, is that many of these are implicit in our physical theorizing, so that they should not be as controversial as they might initially seem.

One overarching theme is that we should take the mathematical structures of our best physical theories *seriously* in telling us about the nature of the physical world. I have discovered many people to balk at this thought, but it seems to me to follow reasonably straightforwardly from a general commitment to scientific realism, to the view that (roughly) our best scientific theories tell us about, or are getting increasingly close to telling us about, the true nature of the world; that these theories are not merely predictive devices or instruments, and this is why they are as successful as they are. This is a view that stands in marked contrast to, for one, Niels Bohr's famous statement that, "It is wrong to think that the task of physics is to find out how nature *is*. Physics concerns what we can *say* about nature" (quoted in Petersen, 1963, 12). I won't attempt to argue for, let alone define, scientific realism here: this will instead be a background assumption. However, notice how naturally the idea of taking the mathematical structures of our best physical theories seriously goes with the realist thought that these theories are getting at the truth about the world. Since these theories are formulated in mathematical terms, it is natural (for the realist) to think that this mathematics somehow represents physical reality, so that we can learn about that reality from the mathematical formalism. In this regard, keep in mind that the mathematical formulation of a theory is not theoretically inert, in the way that the type of ink in which we write down a theory is, but is bound up with the theory's predictive power.

This is not to say that we should naively read off everything about the physical world directly from the mathematics; that we must take every aspect of a theory's formalism to represent genuine features of the physical world; or that any mathematical structure used to state a theory must represent the world by means of a simple, direct correspondence or isomorphism. Accepting the basic thought that we should take the mathematical structures of our best physical theories seriously does not entail such a crude type of realism or reckless method of interpretation. (Thinking it does so may be the reason people balk at the idea.) Any view that takes the mathematics seriously in ways I will suggest is going to have to pay attention to the differences among various mathematical features of a formulation; to distinguish between the mathematical features that represent genuine physical features and those that do not; and to distinguish the mathematical features that directly represent the physical world from those that do so less directly. It may seem an impossible task to try to distinguish among various types of mathematical features in these ways. But this is something that I do, for better or worse, take up here.

Nor does taking the mathematics seriously mean either reifying the mathematical objects used in a theory's formulation, or somehow taking the physical world itself to be a mathematical object. We can be careful to distinguish between the mathematical structures used to formulate our theories and the physical structure of the world, while at the same time taking the former as a guide to the latter. As Tim Maudlin puts it, to attribute "a mathematical structure to physical items" is to say that those items "have some physical features that make them amenable to precise mathematical description in some respects" (2015, 3). It is not to say that the physical items themselves are mathematical ones, or that the relevant mathematical structure is being reified into a physical thing. Nor, again, is it to say that every mathematical feature of a formalism must be directly possessed by the physical things being represented.

Taking the mathematical structures of our best physical theories seriously simply amounts to the following epistemic idea: that these mathematical structures tell us about the physical world; they provide evidence about the nature of the physical world, so that we can learn about the world from these mathematical structures. I will be arguing that familiar examples in which we take this to be the case show that we do implicitly, and reasonably, adhere to this idea in our physical theorizing.

Maudlin suggests something along these lines as the explanation for the "unreasonable effectiveness of mathematics," in Eugene Wigner's phrase, in describing the physical world. Maudlin attributes the effectiveness to the fact that the structure of the mathematics "directly reflect[s] the structure of the physical world" (2015, 4). As he puts it elsewhere: "There is a longstanding puzzle about why mathematics should provide such a powerful language for describing the physical world. The most satisfying possible answer to such a question is: *Because the physical world literally has a mathematical structure*" (2014a, 52; original italics). My own view differs in certain respects—I think that a mathematical formalism can successfully represent the world even while not literally describing it, in Maudlin's sense, one reason being that a formalism can successfully represent the world rather indirectly, in ways we will see—but the thought behind taking the mathematics seriously is in the same spirit. The thought is simply that there must be *some* kind of correspondence between the mathematics in which we formulate our theories and the nature of the physical world, a correspondence that helps explain, on the one hand, the effectiveness of the mathematics in describing the world, and on the other, the success of our inferences about the world on the basis of that mathematics.

If we are going to take the mathematical structures of our physical theories seriously in telling us about the nature of the world, then we will have to be especially careful whenever there seem to be different mathematical structures we can use to formulate a theory. And it is generally thought that there are alternative mathematical formulations available for any given theory, formulations that are regarded as mere notational variants—different ways of stating one and the same physical theory.

Taking the mathematics seriously may recommend a different attitude. The different mathematical formulations may be more like distinct physical theories, which say different things about the nature of the physical world. Hence a theme I will argue for: cases of mere notational variants in physics are harder to come by than people usually think. Before that makes you too uncomfortable, let me hasten to add that cases of genuine underdetermination of theory by evidence are likewise harder to come by than people usually think, for we often have good reason to choose one formulation over another. I will argue in particular that the mathematical structure needed to formulate the dynamical laws, what I call a theory's dynamical structure,[3] is important both to interpreting a given theory, and to choosing among different theories or formulations (all the while allowing that this is not the sole thing to take into account when it comes to theory choice and interpretation).

In arguing for this, I will make use of a particular way of comparing different structures with respect to their relative strengths or amounts, a means of comparison that goes naturally with the notion of structure I have in mind. Using this means of comparison, we will see at least one example of pairs of physical theories that are standardly claimed to be equivalent, yet whose mathematical formulations utilize different amounts of structure. Taking theories' mathematical structures seriously then suggests that these are not wholly equivalent theories, and that we furthermore have reason to choose one over the other. Although this conclusion will be contentious, I argue that it follows from some general principles we familiarly rely on in our physical theorizing. Let me add as well that this is only to say that a similarity in mathematical structure is a necessary condition on the equivalence of physical theories. I will throughout the book be pointing to places where we must rely on more than theories' mathematical structures in order to draw reasonable conclusions about the physics.

Another theme about theory choice and interpretation. Consider the following passage from David Wallace and Christopher Timpson:

> [I]n our view, there is no guide to the ontology of a mathematically formulated theory beyond the mathematical structure of that theory.... But when trying to learn ontological lessons from the theory, one does well to prefer a representation which makes manifest the structure that the theory ascribes to the world.
>
> (Wallace and Timpson, 2010, 702)

[3] I use this term alternately to refer to the mathematical structure presupposed by the (mathematical) formulation of the dynamical laws, and to the physical structure in the world presupposed by those laws.

Wallace and Timpson use this idea to argue against a certain view on the metaphysics of quantum mechanics, known as wavefunction realism, which holds that the mathematical wavefunction used in the quantum formalism represents a real physical field. Against such a view, they claim, "there is a far more perspicuous way to understand the theory" (2010, 701).

I will be defending something similar to Wallace and Timpson's criterion of perspicuousness, though I take the idea a bit farther. (At the same time, we will see a way in which I take it less far.) I will be suggesting that we adhere to a criterion of directness in choosing a mathematical formulation of a physical theory: we should, other things being equal, prefer a formulation that most directly corresponds to the nature of the physical world, for this brings with it a level of "metaphysical perspicuousness" that is preferable. The reason I say this goes farther is that even though Wallace and Timpson suggest that perspicuousness is important, they do not say that the perspicuous formulation is preferable for the reason a typical realist wants to hear. Wallace and Timpson deny that the perspicuous formulation is preferable because it most accurately represents the world—or at least, they aim to remain neutral on this. They are agnostic about whether, when there are competing mathematical formulations that appear to depict different physical realities, one of them must be the correct or most accurate representation, or whether the different mathematical representations may instead be equally correct, even if one is most perspicuous (while not being closest to the truth).

I am going to suggest the stronger position: the more perspicuous formulation, the one that is preferable for that reason, more directly gets at the true nature of physical reality. (That said, figuring out which formulation is most direct is a complicated and subtle business, for reasons that won't be fully articulated until the end of the book.) In fact, I confess to finding the Wallace–Timpson position somewhat puzzling. What do they mean by a perspicuous representation, if not a particularly clear-eyed representation of the true nature of the physical world? Against their agnosticism, I think the more perspicuous representation provides the more accurate description of physical reality, and this is *why* it is more perspicuous. It is for that reason preferable (to the realist), other things being equal. (Although Wallace and Timpson claim to be realists, their view raises the specter of antirealism, in ways I will discuss.)

One way for a formulation to be more direct is by being stated in terms of things that are themselves more directly about the physical world. As a result, by the lights of the directness criterion, theoretical formulations given in terms of reference frames or coordinate systems, although common in physics books and useful for many purposes, are less preferable, other things being equal, because they are less direct. Reference frames and coordinate systems are devices we bring to bear for the purposes of describing physical systems, not inherent in physical systems themselves, and they do not directly characterize their natures. They are indirect, if useful, descriptive tools.

Why is a direct formulation preferable? I say more about this later on, but one thought is the following. Consider Aristotle's physics, according to which certain elements tend to move toward the center of the spherical universe, and other elements tend to move away from it. As a result of these tendencies, certain coordinate systems—namely, those with an origin located at the center of the universe—will be particularly natural or useful for describing systems' behavior. If someone were then to ask *why* those are especially good coordinate systems to use, it would be unsatisfying to refrain from giving an answer. We should be able to give an answer in terms of the nature of the physical reality the theory describes—in this case, by pointing to the fact that there is a dynamically preferred spatial location, which is well-reflected by any coordinate system that has its origin located there. The explanation should not bottom out at a brute reference to coordinate systems, let alone particularly well-suited coordinate systems, but at the nature of the physical reality that *makes* those coordinate systems natural to use. Direct formulations are for analogous reasons more explanatory: we understand more readily what it is about the world that makes the formulation as good and useful as it is. (Compare Frank Arntzenius and Cian Dorr on standard mathematical definitions of differentiable manifolds in terms of coordinate charts. Such definitions are "spectacularly unsatisfying from a foundational point of view" (2012, 232), since they mention things like "admissible coordinate functions" without saying what it is about the intrinsic structure of the space that makes those coordinate functions admissible.)

Another thought in the background is that indirect formulations leave too much room to appeal to anything one likes in formulating the laws, resulting in a type of theory that the realist, at least, won't be happy with—one that seems like a "cheap instrumentalist rip-off," in the memorable phrase of John Earman (1989, 127). Consider the debate between substantivalists, who say that spacetime exists, and relationalists, who deny this. One immediate concern for the relationalist is that the laws of standard physical theories seem to be stated in terms of, or at least to presuppose things about, spacetime and its structures. In reply, the relationalist can say that we should not take these references to spacetime so seriously: although we make use of "spacetime" in formulating the laws, there really is no such thing. According to such a relationalist, the laws are formulated indirectly, in terms of something that is not directly about the physical world, but is nonetheless mentioned for purposes of theorizing about systems' behavior. This feels like cheating. It feels like the relationalist is saying: "things behave as if there were spacetime; we are justified in referring to spacetime in our theory; but [*psssst!*] there really is no such thing"—an instrumentalist rip-off. A direct formulation leaves little room for cheating, and is for that reason preferable.

More generally, directness brings with it a level of perspicuousness that aids the interpretive project of figuring out what physics is saying about the world. It can also yield theoretical progress: classical electromagnetism is a case in point.

Maxwell formulated the equations that go by his name in terms of the potentials, and the result was a bit of a mess. Oliver Heaviside was the one who came up with the improved and streamlined version we are familiar with today. As it happens, Heaviside's formulation is also more direct: the equations are formulated in terms of the electric and magnetic fields, which the theory takes to be genuine physical entities in the world, rather than the potentials, mathematical devices that don't directly correspond to physical things. Heaviside was led to his formulation by thinking carefully about what the theory is saying about the nature of physical reality, and trying to devise a mathematical formulation that directly mirrors that reality.

One unexpected consequence of prizing both directness and perspicuousness is that even though we should generally prefer formulations of physical theories that do not depend on coordinates or other auxiliary descriptive devices—in accord with current thinking in foundational discussions—at the same time we needn't eschew coordinate-based reasoning in physics altogether—against much current thinking. Although the current fashion in foundations of physics is to avoid all mention of coordinates, we will see that there are ways of reasoning about physics by means of coordinates that are useful and legitimate, even perspicuous. Another theme: the role of coordinate systems in physics is more subtle and complicated than usually acknowledged. One distinction to be made in this context is between formulations or claims that are coordinate- or frame-*dependent*, in the sense that they vary with the particular choice of coordinate system or reference frame, and formulations or claims that are what we might call coordinate- or frame-*based*, meaning simply that they mention or refer to coordinate systems or reference frames. Although the former type of claim can be misleading as to the true nature of physical reality—as, say, (frame-dependent) claims about time elapse in special relativity do not get at the underlying nature of the world according to the theory—the latter can be a useful, even straightforward, guide to the nature of physical reality. One example I will emphasize is how we commonly take the mathematical form of the laws in different kinds of coordinate systems or reference frames to indicate the underlying nature of the world, similarly to how the form of the metric in different types of coordinate systems indicates the structure of the Euclidean plane. This kind of feature makes reference to coordinate systems or reference frames, but it does not thereby fail to indicate the underlying nature of (physical or geometrical) reality. On the contrary.

It may be evident by now that I endorse not only a standard or old-fashioned type of realism, but also the "ideal of pristine interpretation," in the phrase of Laura Ruetsche (2011, Ch. 1). Ruetsche explains this ideal, and her reasons for rejecting it, in detail throughout her book, but the basic idea is what we might think of as the standard realist view of theory interpretation: that our best physical theories tell us about what the world is like; that there is one—and only one—way the world is, according to a theory, so that we should "interpret a theory in the same way no matter what conditions befall it" (2011, 343). The pristine ideal includes the further thought that a theory tells us about what other physically possible worlds are like, in ways dictated by the laws plus some general philosophical principles: something else I endorse. There is one place that I part with the pristine ideal as Ruetsche conceives of it, though. Although the interpretation of a theory stems from the laws in accord with some general principles, I do not think that those principles are themselves wholly "antecedent" or held "come what may," as Ruetsche (2011, 4) says of the pristine ideal. (Nor do these principles suffice to pick out an interpretation: some additional physical posits will be required, in ways I will discuss.) Rather, the sorts of principles I will defend are epistemic principles that are both informed and justified by examples of successful theorizing that rely on them; and these principles hold only ceteris paribus. We nonetheless do well to abide by them.

Against the pristine ideal, Ruetsche sees theory interpretation as a more piecemeal, pluralistic, and pragmatic affair. She argues that the pristine alternative is untenable once we look carefully at quantum field theory and the thermodynamic limit of quantum statistical mechanics. I won't be addressing those physical theories here, but will assume the viability of a pristine interpretation for the theories I do discuss: a background assumption. I simply find it too hard to reconcile my realist tendencies with the possibility of "multiple interpretations in the standard sense" (Ruetsche, 2011, 10). Similar considerations lead me to assume a fundamentalism and universalism about the physical laws, against Nancy Cartwright's (1999) view that the world is "dappled," bringing with it a patchwork of laws that accurately describe things in only a piecemeal way: another idea too difficult to reconcile with a thoroughgoing scientific realism. Perhaps I will eventually be forced to some such view on the basis of the physics Ruetsche discusses; until then, I aim to hold onto a more comprehensive realism and see where it gets me. (In fact, I would put things more optimistically, though I won't explore this in detail here: the approach I take and the lessons I draw for the theories I discuss, including the viability of realism, should carry over to any theory we take to be a genuine candidate fundamental theory for our world—so long as it is a bona fide physical theory, with a base of empirical evidence and certain core principles and posits concerning the physical ontology. The reason to focus on the theories I do is that their core principles and mathematical structures are comparatively well-understood and -delineated, making it easier to extract the sorts of lessons I aim to extract.)

That said, there is room for a certain stripe of antirealist to agree with much of what I say in this book. For not all antirealists are instrumentalists or Bohr-type antirealists. And the antirealist who does not go so far as those other views can agree that the mathematical structures of our best physical theories tell us about the nature of the physical world, and can even endorse the interpretive project of trying to figure out what they are telling us about the world, but simply deny that we should believe what they are telling us—an attitude that Ruetsche herself adopts. Bas van Fraassen's constructive empiricism (1980) is another view that would fall into this camp (at least when it comes to theories' claims about the unobservable aspects of the world). Indeed, one might turn the tables to question whether it is possible to maintain the kind of realism I advocate toward the various physical theories I discuss—primarily classical mechanics and spacetime theories, both classical and relativistic, but also classical electromagnetism and non-relativistic quantum mechanics. These theories famously clash in ways that spell trouble for adopting a blanket realism toward them all.

My instincts nonetheless point elsewhere. We can be full-throated realists about these theories severally, treating each as a candidate fundamental theory in its turn (an approach I adopt in this book). We should even be able to be realists about them all (albeit in a way that may require such notoriously tricky notions as relative fundamentality and approximate truth). Once again, it is not my aim to argue for such a position here. I will simply assume a thoroughgoing realism, even if this is not absolutely required by my approach, and even if there is work left to be done in making sense of it. However, I can't complain if what I say turns out to be acceptable to certain antirealists, and I will point to places where it does seem as though such an antirealist can agree with my discussion.

I will also be suggesting that we take what we might call a theory's "metaphysical aspects" seriously: another theme. Valia Allori (2015a) argues that we cannot investigate the invariances or symmetries of a physical theory independently of its metaphysics. I will be defending a more general version of this idea. Although I do argue that the mathematical formalism of a theory is a guide to the nature of the physical world, it is also the case that we cannot get at that nature wholly independently of a theory's metaphysical aspects, which go beyond the formalism. Simply put, not everything about the physical world can be read straight off the mathematics. One reason is that some initial physical posits will invariably play a role as well. To give an example I will return to: what is the world like according to Newtonian gravitation? The answer will depend on whether we take the theory to be about particles with gravitational forces acting on them at a distance, as suggested by the traditional formulation of the theory, or particles whose motions are instead affected by the local spacetime structure, as suggested by the "geometrized" formulation of the theory.[4] Some initial physical assumptions (such as whether to countenance gravitational forces) must be made before we can fully discern the nature of the physical world according to the theory, as well as which mathematical formulation to focus on in this interpretive project in the

[4] Geometrized Newtonian gravity, also called Newton–Cartan theory, was first developed by Élie Cartan and Kurt Friedrichs in the 1920s. It is formulated in terms of the standard mathematical formalism of general relativity, resulting in a theory that appears to "geometrize away gravity" in a manner similar to general relativity. Presentations are in Friedman (1983, Sec. 3.4); Malament (2012, Ch. 4).

first place. The mathematics in which a theory is couched is a guide to the nature of physical reality, in other words, but this is not to say that that nature is completely determined by the mathematics. Metaphysics—or what, for reasons to come, may be better called simply *physics*—matters too. We will see that the mathematical and metaphysical aspects of a physical theory are intertwined in various ways that are not completely straightforward.

More generally, this book will involve some metaphysics of physics. This goes against what Juha Saatsi calls the "well-motivated anti-metaphysical trend in the epistemology of scientific realism" (2019, 146), especially popular among structural realists. Saatsi is in favor of that trend and against any "deep metaphysics" of physics that goes beyond the basic commitments of the realist, which in his view include only an adherence to the truth of our theories insofar as this is needed for predicting and explaining the phenomena. Anything beyond these basic commitments devolves into speculative metaphysics that the realist need not—should not—commit to (something he says is particularly evident in philosophical discussions of quantum mechanics).

Although there will of course be epistemological issues that arise for any realism daring to delve into the metaphysics of physics, to my mind this sort of thing does not go beyond the realist's commitments, but is part and parcel of a basic realism, not to be abandoned in the face of the difficult epistemological questions that result. In particular, it is part and parcel of a realism applied to candidate fundamental physical theories, which are formulated in abstract mathematical terms, and thereby require some metaphysics—or again, just plain physics—in order to paint the picture of the world they describe. In my view, Saatsi's own "minimal realist attitude" simply falls short of realism: it is an untenable stopping point, refusing to dig deeper into the nature of the reality responsible for the empirically confirmed predictions and explanations we get from science. Indeed, the kind of realism that Saatsi endorses would wind up eschewing too much of the science he claims to want to preserve, such as scientific explanations of the phenomena, as I discuss at the end of the book. For these reasons, I do not see my approach as "metaphysical hubris" (in Saatsi's phrase), but a basic part of science as ordinarily understood.

This brings me back to a theme mentioned earlier. Given my emphasis on both the metaphysical and mathematical aspects of physical theories, I will frequently see a non-equivalence between theories or formulations where others see equivalence. For example, I argue that two formulations of classical mechanics that are ordinarily taken to be equivalent (the Lagrangian and Newtonian formulations) differ not only in mathematical structure, but also in various metaphysical respects. Both kinds of differences are significant enough to warrant regarding these as distinct physical theories, with differing accounts of what a classical mechanical world is really like. Or take the Heisenberg and Schrödinger theories or "pictures" of non-relativistic quantum mechanics, generally seen as mere notational variants. Although these are mathematically equivalent in a certain sense, there are plausibly significant metaphysical differences between them. On a natural understanding of the Heisenberg picture, there is one state of the world that is unchanging with time, whereas on a natural understanding of the Schrödinger picture, physical states themselves evolve in time. Taking theories' metaphysical aspects seriously then suggests that these are distinct physical theories, with different accounts of the physical world (subtleties to be elaborated on later). I will be emphasizing these sorts of metaphysical as well as mathematical differences between theories, which will lead me to see more cases of inequivalent theories than many people will be happy with. That said, I can still talk of the various respects in which theories are, or are not, equivalent to one another, and in that way retain what is of value behind standard claims of equivalence in physics.

All of these themes are interrelated, and in my view they are all bound up with a commitment to scientific realism. Again, I will not argue for realism here. Nor will I offer an account of scientific theories, theoretical equivalence, laws of nature, scientific explanation, or fundamentality—even though all of these have a role to play in what follows. What I say can have ramifications for these things, in ways I will discuss. Yet I take it that my discussion can proceed without having to give explicit accounts of these other notions, each of which could take up a book on its own.

A final theme has more to do with philosophical temperament than anything else. A lot of recent philosophy of physics has been marked by a strongly formal turn. This book goes against that trend. Although I will be focusing on the mathematical structures of physical theories, the discussion here contains much less than is typical of recent journal articles in the way of mathematical formulas or proofs of theorems. More generally, I eschew many of the mathematical methods adopted by philosophers of physics these days, even though I do discuss and apply ideas that come directly from mathematics. I simply take a different approach, one that aims to minimize explicit use of mathematics and technicality as much as possible. (As much as possible: occasionally things will unavoidably get more technical; when that is the case, I aim to make the discussion accessible to the uninitiated.) I try to get to the bottom of things in as non-technical, simple and straightforward, a manner as possible. In this I am influenced by my dissertation advisors, David Albert, Barry Loewer, and Tim Maudlin, who demonstrate how much good philosophy can be done in the absence of explicit mathematics. Their work is of course deeply grounded in the mathematics of physics, but the formalism tends to remain in the background, brought to the fore on an as-needed basis. I cannot hope to accomplish a fraction of what they do, but I greatly admire their model of doing philosophy of physics, and try to emulate it.

To put it a little more strongly: although I will be emphasizing the importance of the mathematical structures used to formulate our best physical theories, I also think that one can miss significant aspects of what physics is saying by focusing *too* closely on the mathematical formalism. Hans Halvorson bemoans the "decline in standards of rigor" in recent philosophy of science (recent discussions of theoretical equivalence being a case in point), noting that many a technical term

> has made its way into philosophical discussion but has then lost touch with its technical moorings. The result is almost always that philosophers add to the stock of confusion rather than reducing it. How unfortunate it is that philosophy of science has fallen into this state, given the role we could play as prophets of clarity and logical rigor. (Halvorson, 2019, 10)

Halvorson suggests that bringing the rigor and clarity of logic and mathematics back into philosophical discussion, even to the point of reframing various philosophical questions in explicitly formal or mathematical terms, will serve to clarify many outstanding issues in philosophy of science.

I agree that formal methods can clarify points at issue and that the mathematical turn has produced useful results. But I also think that a laser focus on the mathematical formalism can lead one to miss the forest for the trees, for it can lead one to miss important aspects of the physics. In particular, it can lead one to miss the "picture of the world" presented by a physical theory, in a phrase I use in Chapter 7. As Richard Feynman says, within a discussion emphasizing the usefulness and importance of mathematics to physics,

> Physics is not mathematics, and mathematics is not physics. One helps the other. But in physics you have to have an understanding of the connection of words with the real world. It is necessary at the end to translate what you have figured out into English, into the world, into the blocks of copper and glass that you are going to do the experiments with. Only in that way can you find out whether the consequences are true. (Feynman, 1965, 55–6)

When doing philosophy of physics, the ultimate aim of which is to understand what physics is saying about the world, we should be careful not to overplay the usefulness of mathematical methods and reasoning, as useful as they can of course be. For certain things in philosophy of physics, including the kinds of things I examine here, formalization is not appropriate—even though I readily agree that there are cases in which one gains real insight by formalizing a question in philosophy of physics.

Note too that a lack of formalization does not necessarily mean a lack of rigor, as I aim to illustrate in this book, although recent literature gives the impression of assuming otherwise (as in the above passage from Halvorson). That said, again, there is such a thing as false precision or rigor or over-formalization, when the phenomenon in question does not possess the corresponding level of detail or precision. Consider the epistemic principles that guide our inferences about structure, which I will defend and be relying on throughout. These epistemic rules are not completely precise, nor will they always yield conclusive results. We nonetheless take the inferences they yield to be generally successful, and it is reasonable to rely on them. Such is the way, I submit, with any of our usual criteria of scientific theory choice and interpretation: scientific theorizing is unavoidably messy in this way. And although the philosophical ideal of clarity might seem to demand a decluttering, this is not always feasible or even desirable. I aim to show that there is anyway "rigor enough" in the discussion, in a phrase from Clark Glymour (1977, 236) (quoted at more length in Chapter 3).

The first part of the book (Chapters 2 and 3) discusses the idea of structure I have in mind; how to compare different kinds and amounts of structure; and some general principles governing our inferences about structure, with examples drawn from physics, mathematics, and philosophy of physics. I argue that this idea of structure is familiar, if often implicit, in much of our theorizing, and that the epistemic principles governing our reasoning about it are familiar and generally successful.

The second part of the book applies these ideas to classical mechanics (Chapter 4) and spacetime physics (Chapter 5). In Chapter 4, I discuss the Lagrangian and Newtonian formulations of classical mechanics. I argue that these two formulations differ in structure, in particular their dynamical structure, the structure required by their respective dynamical laws, so that, according to the generally accepted principles from Chapter 3, we should conclude that they are not fully equivalent: they say different things about the fundamental nature of a classical mechanical world, contrary to the usual view that they are mere notational variants. In Chapter 5, I suggest that reformulating the traditional debate between relationalists and substantivalists about spacetime in terms of a notion of spatiotemporal structure serves to reorient the debate so as to render it directly relevant to current physics, against recent claims that this dispute is non-substantive, outmoded, or wholly divorced from physics.

The final part of the book (Chapters 6 and 7) addresses concerns that arise from taking this notion of structure seriously, including the worry that I take the mathematical structures of our physical theories *too* seriously, and related questions having to do with the equivalence of physical theories. It is in Chapter 7 that the metaphysical aspects of a theory will come to the fore as being equally important to its mathematical structure for understanding what the theory is saying about the world. It is here, too, that the entanglement between theories' metaphysical and mathematical aspects will more fully emerge.

I won't be discussing more cutting-edge physics, such as quantum field theory or various programs in quantum gravity. Nor will I even much discuss non-relativistic quantum mechanics, aside from some examples used in Chapter 7. I hope to show that there are nonetheless interesting, important, at times surprising lessons to be had from thinking about more pedestrian physics, lessons that should carry over to other physical theories.

I once overheard a philosopher I admire remark that philosophers should not write books, for in this person's experience, the book never contained anything beyond what had been said by the author, much more succinctly, in previously published papers. I am to some extent guilty of this: some of what I say here is an elaboration on what I had been trying to say in a few earlier papers. Yet in writing those papers, it soon became clear to me that I couldn't say all that I wanted or needed to within the length of a standard philosophy journal article. Referee reports would come back to me with questions and objections that, I felt, I could answer—if only I had the space in which to do so. This book is my attempt to do that.

2

What is Structure? Why Care about It?

> If something is in me which can be called religious then it is the unbounded admiration for the structure of the world so far as our science can reveal it.
>
> Albert Einstein (in a 1954 letter)

2.1 An example

The first order of business is to explain the conception of structure I will be working with and the related idea of comparing different structures. I won't be defining the notion (I don't think it can be defined in more basic terms), but will illustrate it by example. Although I use it in a particular way, the basic idea is not novel. The examples will reveal that it is implicit in many aspects of our theorizing in physics, mathematics, and philosophy of physics.

In this chapter I discuss some examples from mathematics and physics that get at the general idea. I begin with a simple example from mathematics in this section, which will lay the groundwork for locating analogous ideas in physics in Section 2.2. In Chapter 3, I will look in more detail at some familiar inferences about physics that rely on the notion, in order to extract some general principles governing our reasoning about it.

Start with a simple example from mathematics that will illustrate the main ideas. Consider a two-dimensional Euclidean plane. What is the nature of this space?

One way of getting at an answer to this question, though it is somewhat indirect, is to consider the different kinds of coordinate systems we can use. A coordinate system is a device for labeling the points in a space by means of numbers: to each point, it assigns a unique numerical "address." We use these coordinate numbers to represent the space and the points within it. For a two-dimensional space, we use two real numbers to pick out the location of each point. (We do not have to use real numbers—we could use complex numbers or something more abstract—but we may assume this here for simplicity. In general, for an n-dimensional space, a coordinate system will assign each point a unique n-tuple of real numbers, so that a coordinate system is a one-to-one map from the space into $\mathbb{R}^n$. This is roughly

Physics, Structure, and Reality. Jill North, Oxford University Press (2021). © Jill North.
DOI: 10.1093/oso/9780192894106.003.0002

what we mean when we say that a space is n-dimensional: we need at least n independent numbers to uniquely pick out a location.[1])

There are (infinitely) many coordinate systems we can use for the plane, (infinitely) many ways of assigning real numbers to the points. These different coordinate systems will be related to one another mathematically. There are transformation equations that specify the coordinates of a point in one coordinate system, given its coordinates in another. These equations give the coordinates in one coordinate system as functions of each of the point's coordinates in the other. (For a transformation from coordinate system (x, y) to coordinate system (x', y'), the equations have the form $x' = f(x, y)$ and $y' = g(x, y)$.)

From among all the coordinate systems we can use for the plane, there is a particularly nice kind, the Cartesian coordinate systems, which have straight, mutually orthogonal coordinate axes, and whose numerical values reflect the relative locations of the points in a particularly clear manner.[2] Even assuming a Cartesian coordinate system, there remains quite a lot of freedom in which coordinate system to use. We can rotate or translate or reflect different Cartesian coordinate systems relative to one another and still have a perfectly good Cartesian coordinate system for the plane, a way of uniquely labeling each point by means of Cartesian coordinate numbers.

Now think of all the different Cartesian coordinate systems we can use for the plane, and think of the similarities and differences among them. The different coordinate systems will in general disagree on the pair of numbers, the coordinate values x and y, assigned to a given point.[3] As a result, they will also disagree on the differences between the x- or y-coordinate values of different points. Given a Cartesian coordinate system (x, y), if we transform to a new coordinate system (x', y') by means of a translation or reflection or rotation, in general $x_p - x_q = \Delta x \neq \Delta x' = x'_p - x'_q$ and $y_p - y_q = \Delta y \neq \Delta y' = y'_p - y'_q$ for two points p and q.

In an important sense, though, these differences among the different coordinate systems do not matter. Any one of these coordinate systems is a perfectly legitimate way of labeling the points in the plane. Different coordinate systems may disagree on whether a certain point p is assigned x-coordinate value 3 or 4, say, but this

[1] That is only rough. We can code up the information contained in n distinct numbers by means of one number (using decimal expansions), though such a coordinatization would not be very useful. Different definitions of the dimensionality of a space are available in different branches of mathematics. The above assumes that there is a global coordinate system, something that is available for the Euclidean plane. In the most general case we will have to coordinatize the space in local patches, via maps from subsets of the space to subsets of $\mathbb{R}^n$, with compatibility conditions on their overlap.

[2] One further ingredient that I set aside until later in this section: what counts as a Cartesian coordinate system, in the sense I have in mind, also presupposes facts about the metric of the space. For all that I have said so far, standard Minkowski, or Lorentz, coordinates, the standard inertial coordinate systems used in Minkowski spacetime, would count as Cartesian coordinate systems. In one sense, they do qualify as such: these are coordinate systems with straight, mutually orthogonal coordinate axes, whose coordinate values perspicuously reflect points' relative locations. But in another sense, they do not count as truly Cartesian coordinates, in that they do not respect a specifically Euclidean metric. As will be clear by the end of this section, it is this latter sense that I ultimately have in mind.

[3] There may be a fixed point, which will have the same coordinates in each coordinate system (the origin under a simple rotation, for example).

is not a "real disagreement" between them; it is not as though one of them gets the coordinate value *right* and the others get it *wrong*. The points receive different numerical addresses by the different coordinate systems. But the points themselves are unchanged when we transform from one coordinate system to another; they are simply described differently.

The reason these differences among the coordinate systems don't matter is this. There are some things that all these coordinate systems agree on, despite their disagreement on such things as the coordinate values of a given point or the differences between the x or y coordinate values of distinct points. All Cartesian coordinate systems will agree on the distance, $d = \Delta s$, between two points given by the familiar Pythagorean equation, $d^2 = (\Delta s)^2 = \Delta x^2 + \Delta y^2 = \Delta x'^2 + \Delta y'^2$. Rotate or translate or reflect the coordinate system, and this quantity will remain unchanged. (That this is the case can be shown from the transformation equations between Cartesian coordinate systems.) We say that the distance between any two points is *invariant* under, or unchanged by, such changes in coordinates.

(I have been putting things in terms of passive transformations, transformations that alter the coordinate system being used while leaving the objects like points alone. This is as opposed to an active transformation, which transforms the objects while leaving the coordinate system alone. For any passive transformation, there is typically a corresponding active one, and vice versa: the corresponding active transformation will move the objects around in such a way as to yield the same changes to objects' coordinate-dependent descriptions that the passive transformation does. (If a passive transformation shifts the coordinate system two units to the right, say, then the corresponding active transformation shifts the points two units to the left.) Passive transformations are best for current purposes, but we could put things either way. Although there are complications and subtleties involved in trying to draw a clear-cut distinction between the two types of transformation, a "notoriously muddling subject" (Butterfield, 2004, 88), the basic idea suffices for us here.)

The distance between any two points (given a choice of unit) is the same regardless of the coordinate system. This is a *coordinate-independent* feature of the space. Since any of these coordinate systems is an equally legitimate way of labeling the points in the plane, this distance measure, which is agreed upon by all the coordinate systems, seems to be part of the intrinsic, objective nature of this space—an aspect of the plane itself, apart from our descriptions of it. The different coordinate systems simply describe that nature differently, dividing up the distances into different Δx and Δy parts. (I aim to discuss this in a way that does not require delving into the metaphysics of intrinsic versus extrinsic properties, instead relying on the intuitive idea that a feature is intrinsic to an object just in case whether the object has the feature depends solely on what the object itself is like, setting aside any concerns that arise from taking this to be a proposed analysis. The related idea of a feature that does, as opposed to one that does not, depend on our descriptive devices—in the same way that coordinate values do, but distances do not, depend on our choice of coordinate system for the Euclidean plane—should be clear enough.)

Coordinate systems are labeling devices that we impose on a space for purposes of describing it; they are not inherent in the space itself. That is why the things the different coordinate systems disagree about don't matter. Those *coordinate-dependent* features, which vary with the choice of coordinate system, are not about the space as it is in itself, but are about our descriptions of the space. (Rather, as I will emphasize in Section 2.3, they are not *solely* about the space itself, but are *in part* about our descriptions of it.) This, in turn, suggests that the choice of coordinate system is just an arbitrary choice in description, a conventional choice to be made from among equally legitimate representations or descriptions of the space.

I said that the distance between any two points in the Euclidean plane is a coordinate-independent, invariant (unaltered by changes in coordinates) feature of the plane. Yet the Pythagorean equation given above, although agreed upon by different Cartesian coordinate systems, is not wholly independent of coordinates. A distance formula like that one, expressed in terms of coordinates such as x and y, can change depending on the type of coordinate system. If we were to use non-rectangular coordinates, then that formula would not calculate distances correctly. In polar coordinates (r, θ), for example, the distance $d = \Delta s$ between two points with coordinates (r_1, θ_1) and (r_2, θ_2) is given by the more complicated expression $d^2 = (\Delta s)^2 = r_1^2 + r_2^2 - 2r_1 r_2 \cos(\theta_1 - \theta_2)$, which does not have the familiar Pythagorean form.

There is a fully invariant way of expressing distances on the plane, however. This is not given by a coordinate-dependent formula like the Pythagorean equation. It is given by a geometric object called a (metric) tensor, which is a generalization of the idea of a vector. A tensor is an abstract geometric object that is invariant under coordinate changes, just as a vector is. As with vectors, tensors can have different components in different coordinate systems; yet the tensor itself is independent of coordinates, the same object in any coordinate system. In a rectangular coordinate system, the Pythagorean equation correctly gives the components of the metric tensor on the Euclidean plane; in a different kind of coordinate system, it does not. But the Euclidean distance between any two points, this number (in a given unit), or scalar, is coordinate-independent, the same number regardless of the coordinate system. The metric tensor is a kind of generalization of the Pythagorean theorem, allowing us to calculate distances using any kind of coordinates we can lay down on the plane, whether Cartesian or polar or some other.[4]

[4] The metric tensor effectively codes up local deviations away from the Pythagorean theorem. In arbitrary curved spaces or coordinate systems, there will be a tensor field, with a local metric tensor defined at each point.

(A little more detail. In a two-dimensional space coordinatized by x_1 and x_2, the local metric tensor has the general form $ds^2 = g_{11}dx_1^2 + g_{12}dx_1dx_2 + g_{21}dx_2dx_1 + g_{22}dx_2^2$; that is, $\Sigma_{i,j=1}^{n} g_{ij}dx_idx_j$, where n is the number of dimensions of the space and the g_{ij}'s, the components of the metric tensor, are real-valued functions of the coordinates. In Cartesian coordinates in two dimensions ($x_1 = x$ and $x_2 = y$), $g_{11} = g_{22} = 1$ and $g_{12} = g_{21} = 0$, which yields $ds^2 = dx_1^2 + dx_2^2 = dx^2 + dy^2$, an infinitesimal version of the usual Pythagorean equation. In polar coordinates ($x_1 = r$ and $x_2 = \theta$), $g_{11} = 1$ and $g_{12} = g_{21} = 0$, but $g_{22} = r^2$, which yields $ds^2 = dr^2 + r^2d\theta^2$ as the infinitesimal version of the distance formula. In Cartesian coordinates, the local distance measure has the usual Pythagorean form; in polar coordinates, it does not. The components of the metric tensor, the coefficients g_{ij}, are different in the different coordinate systems. But the metric tensor itself (and various quantities we can construct out of it, such as the curvature) is independent of choice of coordinates, in the same way that a vector is independent of coordinates even though its components are not.)

The metric tensor characterizes the geometry of the plane in a way that is independent of coordinates. The different coordinate-based expressions for the distances between points simply represent that geometry in different ways, in terms of different labelings of the points. That said, note for future reference that the function given by the metric tensor has a particularly simple form when expressed in terms of Cartesian coordinates.

As mentioned above, since the distance measure is unaffected by changes in coordinate system, this seems to characterize the intrinsic, objective nature of the space—in other words: its *structure*. Since any of these coordinate systems yields an equally legitimate representation of the plane, any quantity or feature that is agreed upon by all of them, such as the distance between any two points, is plausibly an aspect of the plane itself, apart from our representations of it. It is plausibly part of the plane's genuine structure. Conversely, since the choice of coordinate system is an arbitrary choice made among equally legitimate representations or descriptions of the plane, any feature that varies depending on that choice, such as the particular coordinate values assigned to a point, is plausibly not a feature of the plane itself, not a part of its underlying structure.

There is an important difference, then, between those features of a space that are agreed upon by all the coordinate systems we can use to describe it—the coordinate-independent, invariant features—and the features that vary with the choice of coordinate system—the coordinate-dependent, non-invariant features, which depend on the labeling system we happen to choose. The former get at the true nature, the genuine *structure*, of the space. The latter have to do with our (conventionally or arbitrarily chosen) descriptions of it.

Just as the plane has an intrinsic nature or structure, so too for other kinds of mathematical objects. Consider vectors, which we can think of roughly as arrows, objects with both a magnitude and a direction. Relative to coordinate system (x, y), a vector will have x and y components $\vec{v_x}$ and $\vec{v_y}$, the projections of the vector along the x and y axes, respectively. Relative to a different coordinate system (x', y'), the vector will have different components, $\vec{v_{x'}}$ and $\vec{v_{y'}}$, the projections of the vector along the different coordinate axes. But the object itself, the arrow, which is equal to the sum of its components, is independent of coordinate system. It is simply described differently in the different coordinate systems, in terms of different components (just as the Euclidean distance between two points does not vary with the coordinate system even though Δx and Δy on their own do). It is not as though one coordinate system gets the vector's components *right* and the other gets them *wrong*: these are equally legitimate ways of representing the vector. The vector itself is a coordinate-independent, geometric object, which has a particular magnitude and direction that remain the same regardless of the coordinate system we use to describe it.

Mathematical objects like a vector or a plane have a nature, a structure, that is independent of coordinates. Coordinate systems are labeling devices, tools that we impose on these objects in order to represent them by means of numbers. Since many such descriptive tools can be used, we tend to choose one for reasons of convenience. Given all of this, it may seem as though nothing concerning coordinates can tell us about an object's structure—that the mere mention of coordinates renders a feature coordinate-dependent and therefore not indicative of the nature of the thing itself.

However, the role of coordinates is more subtle than that. Consider that the kinds of coordinate systems we can use for the Euclidean plane tells us about its structure. A defining feature of such a space is that it admits of global Cartesian coordinates: a space is flat and Euclidean just in case there is a coordinate system in which the metric takes the simple Pythagorean form (in which the g_{ij} take the form $g_{ij} = 1$ for $i = j$, $g_{ij} = 0$ for $i \neq j$).[5] There are other kinds of spaces (the surface of a sphere, for example) that do not allow for such coordinate systems. A Euclidean plane can be characterized as the kind of space on which we can lay down global Cartesian coordinates. That this characterization makes reference to coordinates does not interfere with the fact that it specifies the plane's structure. (Keep in mind that, as we can see from the equation for the metric function in polar coordinates, a more complicated form of the metric on its own does not indicate a non-Euclidean geometry: that can be the result of using certain kinds of coordinates. The important consideration is whether there exists any coordinate system in which the metric takes the Pythagorean form. That is what indicates a Euclidean geometry.)

[5] Sometimes "Euclidean" is used to refer to any metric that is positive-definite like this one (for which the distance $d(p, q) = 0$ iff $p = q$), including for spaces that are not flat. I reserve the term for flat Euclidean spaces, those that have a geometry satisfying Euclid's postulates.

So even though any particular choice of coordinate system will invariably involve a degree of arbitrariness, a choice made from among equally legitimate representations, features that have to do with coordinates can still tell us about underlying structure. Characterizing the structure of the Euclidean plane by reference to coordinates—specifying that it is the kind of space on which we can lay down certain kinds of coordinates, those in which the metric takes a certain form—may be a less direct means of doing so than by giving the metric tensor, for the simple reason that coordinates are numerical labels we lay down on top of the plane, and there are a variety of different such labelings we can use. The metric tensor more directly encapsulates the geometry of the plane, without any invocation of coordinate labels. Nonetheless, the description in terms of coordinates, while more roundabout, does tell us about the plane's structure. Michael Friedman (1983, Ch. 1) reminds us that it is important to distinguish between the intrinsic features of a space, which are independent of any particular coordinatization, and the extrinsic features, which depend on the coordinatization; for example, to distinguish between the metric tensor and its components in a particular coordinate system. Taking features that reference coordinates—and which are in *that* sense coordinate-based, not in the sense of varying with the coordinate system—as epistemic guides to a space's structure is not to obliterate that distinction. Notice the distinction between a feature or description's being coordinate-*based*, in the sense of mentioning or involving coordinates, and being coordinate-*dependent*, in the sense of varying with or depending on the particular choice of coordinates. A coordinate-independent structure can be characterized by a description that is coordinate-based in this sense, if not coordinate-dependent.[6]

As we see in the case of the Euclidean plane, the kinds of coordinates we can use for a space can indicate its structure. At the same time, that structure indicates what kinds of coordinates we can use. This is not a vicious circle. There are simply two ways of characterizing a given structure, and two corresponding routes to learning about it. A structure can be characterized more directly, as in the case of the Euclidean plane and the metric tensor. Or it can be characterized less directly, by means of the coordinate systems we can use for the space and the features that are invariant under transformations of them.

A further note on coordinate systems. I have been discussing the features of a space that are agreed upon by all the coordinate systems we can use to describe it. Such coordinate systems are often referred to as the "allowable" or "legitimate" or "admissible" coordinate systems. These adjectives make it sound as though any other coordinate system is flat-out illegitimate or disallowed—prohibited. But that can't be exactly right. In *some* sense, we can use any coordinates we like to describe

[6] I hesitate to further muddy the waters at this point, but we will see (for instance at the end of Section 2.3 and in Section 3.5) that even certain coordinate-dependent things can tell us about underlying structure, so long as we are sufficiently careful.

any space or object.[7] Coordinates are just labels, and we can choose to label things however we want: in that sense any coordinate system is allowable. A coordinate system that is stretched in some places relative to an ordinary Cartesian coordinate system might seem prohibited for the Euclidean plane, for instance, since distances calculated in terms of the non-uniformly stretched coordinates won't be the same as in the original coordinate system: the metric structure is not preserved by this transformation. However, we *could* still use such a coordinate system, in that it will label the points uniquely. We could even calculate distances with it, so long as we are extra careful, although the calculations would be unnecessarily convoluted, having to proceed by way of the original coordinate system. When we say that this is not an allowable or legitimate coordinate system, we are saying that we in effect already know what the intrinsic structure of the space is, and that the coordinate numbers would not yield reliable information about that structure. Taking the coordinates of the transformed coordinate system at face value, holding fixed the original underlying distance facts, would be a misleading guide to the structure. An "allowable" or "admissible" or "legitimate" coordinate system is one that sufficiently respects a given structure, even though, in a certain sense, we can always use other ones. (A polar coordinate system is allowable in this sense. The metric function is not as simple as it is in Cartesian coordinates, but distances are still reasonably clear functions of the coordinates. It respects the metric structure (and other lower levels of structure: Section 2.4) well enough.)

Any coordinate system that is a one-to-one function of the points is acceptable in that it will attach a unique label to each point (where this may happen by means of local coordinate charts that together cover the space). Although continuity, for example, is often assumed to be required of any admissible coordinate system, we *could* use coordinates whose values do not vary continuously with the locations of the points. (Continuity is assumed for the kind of space I have been taking for granted, that is, which has the underlying structure of a topological manifold; more in Section 2.4.[8]) As Wallace puts it, "nothing (beyond the fact that life is short) prohibits us from 'coordinatising' a manifold via a wildly discontinuous map from the manifold into $\mathbb{R}$, if we so choose" (2019, 130).[9] That said, such coordinate systems really would not be very useful labeling devices. One particularly useful aspect of coordinates is that they allow us to describe continuous curves by means of continuous functions. Better to use coordinate systems whose continuity matches the continuity structure—the topology—of the space. Indeed,

[7] Subject to existence conditions, that is. We can't use a global Cartesian coordinate system on the surface of a sphere, say, since such coordinates simply do not exist. Sometimes "allowable" or "admissible" does seem to mean only that. The discussion above pertains to a different idea that is often implied.

[8] Occasional exceptions to the continuity requirement—coordinate singularities—are tolerated, as in the coordinate singularity at $r = 0$ and $\theta = 0$ in polar coordinates on the plane.

[9] For example, in the case of a four-dimensional spacetime manifold, "say by using an ordinary coordinate system and then applying one of the standard maps from $\mathbb{R}^4$ onto $\mathbb{R}$" (Wallace, 2019, 130).

the standard definition of a coordinate system in differential geometry builds in this requirement. (The standard definition of a local chart or coordinate patch requires that it respect the local topology. The definition of a smooth atlas, which smoothly stitches together the different local coordinate patches, then adds the requirement that the atlas must everywhere respect the local topology.)

Notice that among the coordinate systems that are allowable in the sense of respecting a given structure well enough, some may be particularly well-suited or well-adapted to it, cohering or meshing with the structure in an especially natural way, and which will in that sense be privileged or preferred. Although polar coordinate systems are legitimate for the Euclidean plane, for example, Cartesian coordinate systems, in which the metric takes a certain simple form, respect the structure better, more naturally or perspicuously. Polar coordinates are perfectly legitimate; nothing goes awry if we use them judiciously; there is even a reasonably straightforward way to calculate distances using them. Yet they do not give as immediate, clear-cut information about the metric structure. The distance formula is more complicated. (Nor do polar coordinates mesh as well with the plane's affine or "straight-line" structure. Some curves that are straight according to the plane's intrinsic structure will be represented by non-linear functions of polar coordinates, and some curves that are not straight according to the structure will be represented by linear functions.) In this sense, Cartesian coordinates are privileged or preferred for the plane: they mesh with its structure in a particularly natural, especially perspicuous way.

There are then two things to keep in mind. First, what is an "allowable" coordinate system depends on the structure we have in mind. Each point in the space should get a unique coordinate assignment, but what is considered an allowable coordinate system typically does more than that, respecting well enough an assumed type of structure; even so, other coordinate systems *can* always be used. Second, there is a difference between saying that various coordinate systems are all equally legitimate or allowable for a space (respecting a given structure well enough) and saying that they are all equally natural or well-adapted to it (meshing with the structure in a particularly natural way)—as polar coordinates are legitimate for the Euclidean plane but are not as well-adapted to it as Cartesian coordinates.

This may seem like a lot of unnecessary detail about coordinate systems. But these things are not usually explicitly mentioned, and this can engender confusion about what, exactly, coordinate systems can, and what they cannot, tell us about underlying structure. We will see this at various points throughout the book.

The upshot of the example of the Euclidean plane is this. There are a variety of different coordinate systems we can use for the plane, all of which are legitimate ways of representing the plane by means of numbers. This suggests, first, that which coordinate system we use is a matter of conventional or arbitrary choice (even though, if we wish to respect a certain aspect of the structure particularly well, the choice won't be completely arbitrary; more on this in Sections 2.2 and 3.5). Second, the features or quantities that are agreed upon by all the different coordinate systems we can use for the plane, the coordinate-independent, invariant features, correspond to the intrinsic nature of the plane, to aspects of the plane itself, apart from our descriptions of it—that is, to what I have been calling its *structure*. In the case of the Euclidean plane, this structure is given by the distance measure (and other lower-level structures, discussed in Section 2.4). As a more general point, there is an important difference between the structure of an object, and the features it has because of how we choose to describe it.

Notice how the fact that many different coordinate systems *are* all equally legitimate ways of describing the Euclidean plane suggests that the plane *has* an intrinsic, coordinate-independent nature, of which these are all equally legitimate descriptions. It naturally suggests that the different coordinate systems are just different ways of describing the very same structure, which is out there in the space regardless of how we choose to describe it by means of coordinate labels. Despite the naturalness of this thought, however, there are other views available. According to a conventionalist like Reichenbach (1958), there is no such thing as "the" structure or geometry of a space, of which there are different allowable descriptions: the structure is as much a matter of arbitrary or conventional choice as the coordinate system is. Another alternative is to maintain that there is an objective structure which the different coordinate systems are all equally legitimate ways of describing, but to deny that we can say any more about what that structure is, beyond giving the different coordinate descriptions and stipulating that they are all equally legitimate; we need not specify (nor perhaps need there even be) an intrinsic nature underlying them all. Such a view does not recognize a distinction between more and less direct characterizations of a space and its structure. This is the view of Wallace and Timpson (2010) (also Wallace, 2012) mentioned in Chapter 1, at least when it comes to physical theories and spaces. (Relatedly, Wallace (2019) argues that not only are coordinate-based and intrinsic characterizations of a given structure equally legitimate, but that neither has primacy over the other.) I return to these alternatives in Sections 5.3 and 6.3.

2.2 Examples from physics

A similar notion of structure is at work in physics. The examples in this section reveal that the same basic idea, along with various details surrounding it, apply just as much in physics. In particular: just as the different coordinate systems we can use for the Euclidean plane tell us about the plane's structure, so too in physics the different coordinate systems, reference frames, or other descriptive devices we use can tell us about underlying structure, in this case the structure of the physical world.

Consider units of measure, one kind of descriptive device we are familiar with using in physics. Spatial distances can be given in terms of feet or meters or some other unit, and the physics will be the same regardless. We conclude from this that physics does not prefer one unit of length over any other, and we may choose any one we like for reasons of convenience. We further conclude that any feature depending on that choice, such as the particular numerical value assigned to the spatial separation between two locations, is not out there in the world apart from a choice of unit. Temperatures can likewise be given in terms of the Fahrenheit or Celsius or Kelvin scale. Nothing in the physics changes when we switch from one scale to another: any of these describe the temperature facts equally well. We conclude that the choice of scale is an arbitrary choice in description, and that any feature that depends on that choice, such as whether one object is twice as hot as another, is scale-dependent, not out there in the world apart from a choice of scale.

Since physics does not recognize or pay attention to differences in scale or unit of measure, since the physics is the same regardless, we infer that the choice of unit or scale is merely an arbitrary or conventional choice in description. There is no "unit of measure structure" in the world. Different units or scales simply provide different, equally legitimate ways of describing the world.

Reference frames are another kind of descriptive device we commonly use in physics. Think of a reference frame as a certain kind of coordinate system, one that is attached to an observer, representing the observer's own point of view. (It need not be attached to any actual observer. For our purposes, we may treat reference frames and coordinate systems interchangeably.[10]) A physical theory will typically be invariant under particular changes in reference frame, in that the laws will remain the same under such changes in frame.

Newton's laws, for example, are invariant under changes in inertial reference frame—transformations from the coordinates of one reference frame moving with constant velocity to the coordinates of any other frame moving with constant velocity relative to the first. The laws "say the same thing," they make the same predictions, regardless of which inertial reference frame we use for describing a system. (Put another way, the laws have the same mathematical form in any inertial frame. More on this way of putting it in Chapter 3.)[11] Nothing of physical significance changes when we choose a different inertial frame: as far as Newton's laws are concerned, any such frame yields an equally good description of things. We conclude that the choice of inertial frame is an arbitrary choice in description.

[10] Even though, as Norton (1993a, Sec. 6.3) emphasizes, these are conceptually distinct.

[11] That is, when expressed in terms of the coordinates of any inertial frame, the laws always yield the same predictions, and in this sense they say the same thing regardless of inertial frame: any observer, describing things in terms of the coordinates of any inertial frame, will make observations that confirm these laws. That said, in some sense, the laws can yield the right predictions when stated in terms of the coordinates of any reference frame, analogous to how we can use any coordinate system to describe a given mathematical space or structure. We need the idea of an equation's form to make the point clearer, more on which later in this section and in Chapter 3.

We further conclude that any quantity that depends on this choice, like velocity, is frame-dependent, not out there in the world apart from that choice. In other words, we conclude that there is no "absolute velocity structure" in the world, no fact about what velocity an object *really* has. An object's velocity depends on the inertial frame we use to describe it, and any such frame is equally legitimate as far as Newton's laws are concerned. (This is according to a familiar conception of the theory. Newton's own conception disagrees on some of these points, as we will see in the next chapter. Mind you, the invariance of the laws under transformations of inertial frame is not in question; where Newton disagrees concerns the structure underlying these laws.)

The laws of special relativity (Maxwell's equations and the Lorentz force law) are likewise invariant under changes in inertial frame, in this case changes in Lorentz frame, so-called because the different inertial frames are related by the Lorentz transformation equations,[12] unlike the inertial frames of Newtonian physics, which are related by the Galilean transformations. (Lorentz frames are still inertial frames, but the spacetime structure differs from Newtonian physics, and this picks out a different class of reference frames as inertial.) These laws "say the same thing," they make the same predictions, regardless of which Lorentz frame we choose. (They have the same mathematical form regardless.) Since the physics does not recognize or pay attention to differences in Lorentz frame, we infer that the choice of such a frame is merely an arbitrary choice in description, a conventional choice to be made from among equally legitimate representations. We likewise conclude that any quantity that depends on this choice, such as the temporal separation between events, is frame-dependent, not out there in the world apart from that choice. There is no "absolute simultaneity structure" in the world, no fact about which events are *really* simultaneous with one another. The simultaneity of (spacelike separated) events depends on the choice of Lorentz frame, any one of which is equally legitimate according to this physics. (This is according to a familiar conception of the theory. A Lorentzian conception differs on some of these points, as we will see in Chapter 3. Mind you, here as well the invariance of the laws is not in question, but rather the structure underlying these laws.)

On the other hand, any feature or quantity that the different reference frames or units of measure *do* all agree on—any quantity that is in this way independent of our arbitrary choices in description, being ascribed to a system regardless of which choice we make—does seem to correspond to the objective, intrinsic nature of the thing—to its *structure*—in the same way that the distance measure corresponds to the structure of the Euclidean plane.

For example, in Newtonian physics, acceleration is invariant under changes in inertial frame. A system will have the same acceleration regardless of which inertial

[12] More generally, the Poincaré transformations.

frame we use to describe it. Since the laws indicate that the choice of inertial frame is a conventional choice in description, and since a system's acceleration is agreed upon by all the different inertial frames, we infer that this is an objective, frame-independent quantity, something that a system has regardless of how we choose to describe it. In other words, there is an "absolute acceleration structure," an inertial structure, out there in the world: there are facts about which objects are really accelerating or moving non-inertially. Similarly for the times at which events occur: since the different inertial frames all agree on which sets of events are simultaneous, we infer that (unlike in special relativity) there is an absolute simultaneity structure in the world.

In special relativity, the spacetime interval between events, the spatiotemporal "distance"[13] or separation between them, is the same in any Lorentz frame. Since the choice of Lorentz frame is an arbitrary choice in description, and since the spacetime interval between events is the same regardless of that choice, we conclude that this quantity is part of the objective, intrinsic nature of the world, according to the theory. We infer that the spacetime structure of a special relativistic world is Minkowskian, the kind of spacetime that's characterized by this interval. Different inertial frames simply describe this structure differently, dividing up the spacetime interval into different temporal and spatial separations (analogously to how different Cartesian coordinate systems divide up Euclidean spatial distances into different separations between x and y coordinates).

In physics, too, then, there is an important difference between the features that depend on the particular choice of descriptive device, and those that do not. What I am calling structure in physics concerns the latter type of feature: the features or quantities or facts that are agreed upon by all the different descriptions we can use. Since no matter which description we choose we find the same feature or quantity or fact, these things are plausibly out there in the world apart from any of our descriptions of it.

Notice how the very idea of structure, and the thought that there *is* some structure to the world that can be described in any number of different, equally legitimate ways, runs counter to the antirealism of Bohr's mentioned in Chapter 1. Bohr says that our physical theorizing is not about nature itself but about "what we can say about nature." Structure, by contrast, has to do with what all the different things we can say about nature have in common. The fact that there *are* features the different things we can say about the world all agree on suggests that there is a world that has those features, which underlies the different things we can say about it, and which our physical theories are about.

As with the structure of the Euclidean plane, so too the structure of a physical space or object can be characterized either more or less directly. Consider the

[13] The spacetime interval does not strictly speaking yield a distance measure, since it does not satisfy all the conditions of a metric function. It is rather a pseudo-metric.

spacetime structure of special relativity. All inertial frames agree on the value of the spacetime interval between any two events (up to choice of unit). The interval, mathematically specified by a metric tensor, directly characterizes the spacetime structure. Now, in addition, the interval will have the same form when expressed in terms of the Lorentz coordinates (x, y, z, t) and (x', y', z', t') of any two inertial frames: $I = (\Delta s)^2 = -(c\Delta t)^2 + (\Delta x)^2 + (\Delta y)^2 + (\Delta z)^2 = -(c\Delta t')^2 + (\Delta x')^2 + (\Delta y')^2 + (\Delta z')^2$.[14] This gives rise to an indirect characterization of the spacetime structure. The spacetime of special relativity can be characterized by the fact that there exist global Lorentz coordinate systems in which the interval takes this form (even though we can use other coordinate systems, in which the interval takes a different form). This is analogous to how different coordinate systems on the Euclidean plane will agree on the distance between any two points, yet only in Cartesian coordinates will the distance function additionally take the simple Pythagorean form, giving rise to an indirect characterization of the structure in terms of the existence of coordinate systems in which the metric takes that form. The spacetime structure of special relativity can be characterized more directly, with no mention of coordinates, but the indirect characterization also specifies it.

A comment similar to the one about allowable coordinate systems in mathematics applies here too. In physics one often hears mention of reference frames or coordinate systems that are "admissible" or "legitimate" or "allowed" for a given theory. Thus, in Newtonian physics it is often said that only inertial reference frames are allowed; for these are the frames in which Newton's laws hold—in which a particle with no net external force on it travels with constant velocity, for example.[15] Choose a non-inertial reference frame, and it can appear as though an object accelerates for no reason, in particular not because of any force exerted on it by another massive object. This makes it seem as though the use of non-inertial reference frames is truly prohibited by the theory: in such frames, things appear to happen that the theory itself outlaws.

Once again, though, this can't be exactly right. A reference frame is like a coordinate system in being a descriptive device we impose upon systems for the purpose of describing them, not intrinsic to physical systems themselves. In this respect, we can choose any reference frame we like to describe things: we can choose to label things however we want. Indeed, claims of their being disallowed notwithstanding, non-inertial reference frames *can* be used in Newtonian physics, we are *allowed* to use them—indeed we often *do* use them. It is

[14] It is conventional whether the spatial components are positive and the temporal one negative, or vice versa, although it is not conventional that there are three components with one sign and one component with the opposite sign. It is also conventional whether the interval is taken to be I above or instead the square root of this quantity.

[15] Emery (2019) helpfully calls these the "nomological reference frames." However, I would add that these are the frames not only in which the laws hold, but in which they take a particularly simple or natural form, for reasons immediately below and in Chapter 3.

just that the equations expressing the laws in such frames will contain additional "pseudo force" or "fictitious force" terms.[16] These are called "force" terms because they appear to refer to forces, holding fixed the second law's usual connection between force and acceleration. But they are also called "pseudo" or "fictitious" because the things they refer to, if they existed, would not obey the usual Newtonian laws; for that reason, we take them to not correspond to genuine physical forces. We are allowed to use non-inertial reference frames, in other words, so long as we keep in mind that the appearance of extra terms in the equations is misleading, an artifact of having chosen a non-inertial reference frame, rather than an indicator of the existence of physical things that disobey the laws. Non-inertial reference frames are disallowed in Newtonian physics, not in the sense that we are truly prohibited from using them (on pain of falsifying the theory), but in the sense that the usual, simple form of the laws is not preserved. (I say more about this in Chapter 3, with additional subtleties to come in Chapter 7.)

As in the case of coordinate systems in mathematics, therefore, being an allowable type of coordinate system or reference frame in physics does not mean that any other kind is outright prohibited, but that the given type of reference frame or coordinate system respects a particular feature we have in mind, such as a certain kind of structure or a certain form of the laws. Also, as in the mathematical case, here, too, from among the reference frames or coordinate systems that are allowable in the sense of respecting some feature well enough, there may be certain ones that respect it particularly well, and which are thereby privileged or preferred or especially natural. There may be coordinate systems in which the equations expressing the laws take a particularly simple form, for example, as in the case of inertial Cartesian coordinates for Newton's laws, or in which the spacetime metric takes a particularly simple form, as in the case of Lorentz coordinates in special relativity. We can (and often do) use other reference frames or coordinate systems; the physical content of the laws remains the same regardless; nonetheless, certain ones may be preferred or privileged or natural, given the simplicity of expression they give rise to. In this sense, one kind of reference frame or coordinate system is privileged, preferred, or especially natural for the physics, even though in another sense, the choice of reference frame or coordinate system is arbitrary.[17]

[16] Also sometimes called (even more confusingly!) "inertial force" terms.

[17] The idea that there are preferred frames or coordinates in that they yield a simple form of the laws, while at the same time the choice of frame or coordinates is arbitrary, is occasionally mentioned in physics books. One book notes of Newtonian mechanics that, "the first and second laws implicitly imply and require the existence of a certain class of *preferred* Cartesian frames of reference. That Newton's second law and the law of inertia single out preferred frames of reference, called inertial frames, can easily be understood by making a coordinate transformation" to a non-inertial frame, in which additional inertial terms appear (McCauley, 1997, 31; original italics). Another book notes that, "The arbitrariness of the coordinates can be a difficult point for students to grasp because in almost all elementary parts of physics there are a few coordinate systems that are preferred because they make the laws look simpler. For example, there is the class of inertial frames, in which the general laws of special relativistic mechanics take a simple form" (Hartle, 2003, 135). More in Section 3.5.

Since the different choices of descriptive device (coordinate system, reference frame, unit of measure) yield equally good representations of things as far as the physics is concerned, we conclude that any choice we do make is arbitrary or conventional. We further conclude that any feature, quantity, or fact that depends on this choice in the sense of varying with the particular choice we make—as an object's velocity in Newtonian physics varies depending on which inertial frame we use—is not part of the genuine nature of the world, but involves our descriptions of the world. As Marc Lange says of reference frames in particular: "invariant quantities are features of the world, uncontaminated by the reference frame from which the world is being described, whereas frame-dependent quantities reflect not only the world, but also the chosen reference frame" (2017, 142). Structural features, which are agreed upon by all the different descriptions we can use, correspond to features of the world that are "uncontaminated" by any particular representation of it.

Structure, as I understand it, and as it is often tacitly understood in physics and mathematics, concerns the invariant, description-independent features or quantities or facts, those that are the same regardless of choice of description or representation, and which in that way do not depend on our arbitrary or conventional choices of description. Structure has to do with the intrinsic, genuine, objective features or quantities or facts. By contrast, features or quantities or facts that depend on our arbitrary choices in description, those that are not agreed upon by all the different representations we can use, are not wholly about things in themselves, but are in part about our descriptions of things.

That is why we are interested in structure. We want to figure out what the world is really like, according to our best physical theories. We want to reach beyond our representations or descriptions of the world to learn about the nature of the world as it is in itself. We want to "distinguish what is genuinely an aspect of reality from what is a kind of appearance, or artifact, of the particular perspective from which we regard reality" (Price, 1996, 4); to distinguish the genuine structure of the world from the features specific to an arbitrarily chosen representation of it. We wish to learn about the "the structure of the world so far as our science can reveal it," as Einstein puts it in a letter quoted in the epigraph to this chapter.

2.3 Related notions

Structure concerns the features or facts or quantities that are the same regardless of what descriptive device we use, and which, for that reason, plausibly capture the nature of things in themselves. Although there clearly are connections between this idea and other notions—objectivity and invariance, also symmetry, meaningfulness, reality—the connections should not be drawn too tightly. Structure is closely related to these other ideas, but it is not exactly the same as any of them. Without delving into the vast literature on these other ideas, in this section I point to some general ways in which they differ from the notion of structure I have in mind. The differences suggest that we cannot define the notion of structure in terms of any of the others. (This ultimately depends on how one wants to define these other notions, which I don't take a stand on here. I do not conclusively rule it out that a sufficiently elaborated such notion can be made to correspond more closely to that of structure.)

Consider invariance, objectivity, and symmetry. Since structure has to do with features that remain the same under allowable changes in description, structure has something to do with invariance and invariant quantities. This, in turn, suggests that structure has something to do with objectivity. Since the structural features are agreed upon by all the allowable descriptions, and since which description we do choose is a matter of subjective choice, it seems as though the structural features are those that are objective in the sense of being independent of any subjective choice in description. By contrast, features that depend on something like the choice of reference frame or coordinate system seem to be subjective, an artifact of a particular representation or perspective on things. Finally, a symmetry of an object is something you can do to the object so that it looks the same afterward, as a circle is unchanged by a rotation about any angle; we then say that the object is symmetric with respect to that operation. This suggests a close connection between structure and symmetry.

There is a distinguished tradition of drawing connections among these three notions (if not structure per se), including ideas that sound a lot like what I say about structure.[18] Hermann Weyl famously said that, "objectivity means invariance with respect to the group of automorphisms" (1952b, 132). (The automorphisms essentially correspond to the symmetries; mathematically, these transformations form a group, so that we can talk of the features that are invariant with respect to a particular group of transformations, or the group of automorphisms.) Weyl suggests that the features that remain constant under various changes in description are those that get at the nature of the thing in question, independent of our descriptions, and which are therefore its objective features. Katherine Brading and Elena Castellani note that, "It is widely agreed that there is a close connection between symmetry and objectivity, the starting point . . . being provided by spacetime symmetries: the laws by means of which we describe the evolution of physical systems have an objective validity because they are the same for all observers" (2003, 15). Robert Nozick (1998, 2001, Ch. 2) explicitly equates objectivity with invariance, claiming that, "An objective fact is invariant under various transformations" (1998, 21). He discusses the case of temperature, noting that since different choices of scale are equally legitimate, we cannot say that it is

[18] One book-length discussion of various connections among these (taking a different approach from my own) is Debs and Redhead (2007).

"twice as hot" outside when it is 80 degrees as when it is 40 degrees, since this depends on which among the equally legitimate choices of scale we make: this is not an objective feature of temperature. Similarly for the difference between two temperatures $T_1 - T_2$, whereas the ratio between two temperature intervals $(T_1-T_2)/(T_3-T_4)$ is invariant under changes in scale, and is therefore independent of any particular choice of scale. The ratio of intervals is an objective aspect of temperature; temperature differences are not.

All of this sounds a lot like my talk of structure. However, the notion of structure does not exactly overlap with any of these other ideas.

Take objectivity. There can be facts or features that are in a sense objective, and yet are not about structure in the sense I have in mind. This will depend on exactly what one means by "objective," but assume a rough idea of a fact or feature that holds independently of any individual subject or observer or perspective. In this sense, it is not an objective fact about temperature that 80 degrees is twice as hot as 40 degrees, for this depends on which temperature scale we choose, and different observers may legitimately choose different ones. However, it *is* an objective fact, in this sense, that 80 degrees is twice as hot as 40 degrees *according to a particular choice of scale*. This fact holds independently of any particular subject or observer or perspective; it is something that all subjects or observers will agree on. There are similarly no objective, frame-independent facts about velocity in Newtonian mechanics: the choice of frame is a subjective choice in description, and an object's velocity depends on that choice. However, relative to a particular inertial frame S, there will be an objective fact about an object's velocity: any observer will agree that the object in question has that velocity *according to reference frame* S. Likewise, the choice of coordinate origin is subjective; but all hands can agree on the location of the origin given a particular choice of coordinate system.

All subjects or observers can agree that, relative to a particular choice of scale, reference frame, or coordinate system, the fact in question holds. In that sense the fact is objective. Yet such facts are intuitively not getting at structure. Structure has to do with the nature of things in themselves, apart from our arbitrary choices in descriptions of them, and the above facts make essential reference to a particular such choice. Features such as "being twice as hot according to temperature scale T" or "having velocity v in reference frame S" or "having origin $p=(0,0)$ in coordinate system C" make reference to, and so are partly about, an arbitrary choice in description, a particular choice from among all the units of measure or reference frames or coordinate systems we can use. Indeed, it is only by explicitly referring to a particular such choice that the fact becomes objective in the sense of being agreed upon by all subjects. (Recall that our descriptive devices can be mentioned in characterizing structure, as the existence of Cartesian coordinates in which the metric takes a certain form characterizes the structure of the Euclidean plane. But the way in which they are mentioned differs from how they factor into the above objective facts. The characterization of structure refers to the existence of a general type of coordinate system, not any one particular choice; this makes the characterization indirect, not non-structural.)

You might want to consider such facts or features structural simply by virtue of being agreed upon by all observers, thereby bringing structure closer in line with a notion of objectivity. This is to some extent terminological. However, it sounds odd to say that a Newtonian world has a "velocity relative to a particular choice of inertial frame structure" or that a Euclidean plane possesses a "preferred point according to a certain choice of coordinate system structure," suggesting that this is not what we ordinarily mean by structure in these contexts. (This will depend on exactly what one means by "objective," though I suspect that something like this will be the case regardless. Nozick, who argues that the objective facts are those that are invariant under admissible transformations, mentions other aspects of our ordinary notion of objectivity: a fact that is accessible in different ways or among different observers; a fact about which there can be intersubjective agreement; a fact that holds independently of people's subjective thoughts and desires and so on. Facts or quantities or features that mention a particular choice of coordinate system, reference frame, or scale satisfy these criteria of objectivity too, yet intuitively do not get at structure.)

Next consider symmetry. Intuitively and informally, a symmetry is a transformation or (one-to-one) mapping of a structured object onto itself that leaves the (structure of the) object unchanged;[19] we then say that the object is symmetric in that respect or under that operation, as for example a circle is symmetric under rotations. This too sounds similar to my idea of structure, but again the notions are not exactly the same. A symmetry is a mapping or transformation that preserves the structure: it is related to structure, it is an indicator of structure, but is not itself the structure. It is true that symmetries are a particularly important indicator of structure, since structure has to do with the features that are out there in the world apart from our descriptions of it, and one way to discover what those features are is to consider which ones are ascribed to the world regardless of any particular choice of description of it—to figure out which features are agreed upon by all the different descriptions, which can then be said to be symmetric under changes in description. They are nonetheless distinct notions.[20] Symmetry can furthermore mislead us regarding structure, for reasons we will see in Section 3.4.[21]

[19] "Object" in an abstract mathematical sense; the above is not intended to mean that the function is only defined on a single object.

[20] It may seem otherwise on a Kleinian conception of geometry (Section 2.4), according to which we identify a geometry via the invariant quantities under the relevant group of transformations. But even here there is a difference in that the geometry is equated with the invariant quantities, with the symmetry group indicating what the invariant quantities are.

[21] Exactly what is the relationship between symmetries and structure is not something I will explore here. Some recent explorations include the papers in Brading and Castellani (2003); Baker (2010); Belot (2013); Dasgupta (2016); Barrett (2018).

Invariance may be the notion that comes closest to structure, since the invariant features or quantities or facts are those that are unchanged by allowable transformations, or allowable changes in description. Yet these, too, are not exactly the same idea. Again consider a fact such as "object o has velocity v relative to inertial frame S" in Newtonian mechanics. This fact is invariant under a transformation to a different reference frame: it remains true, in any other frame S', that object o has velocity v in reference frame S. Intuitively, however, this is not part of the structure of a Newtonian world in the sense we usually mean in physics. Again by way of comparison, consider how odd it sounds to say that "point p has x-coordinate 3 in coordinate system C" is part of the genuine structure of the Euclidean plane, even though this fact is unchanged by a transformation to another coordinate system. A particular arbitrary choice of descriptive device can factor into a feature that is invariant, yet intuitively the feature will not count as part of the structure.

There is often said to be a link between symmetry, invariance, frame- or coordinate-independence, and/or structure, on the one hand, and what is real or meaningful, on the other. Robert Geroch (1978) distinguishes the quantities or features that "make sense" according to a given spacetime structure from those that do not, the latter being ill-defined. Earman describes how we can compare different spacetime structures by means of which "questions about motion become meaningful" (1989, 36), as it may be meaningful to ask what velocity an object has in one spacetime structure but not in another. Lange says that in special relativity, "mass is a real property (since it is Lorentz invariant)," whereas "total energy and total momentum are frame dependent and therefore not real" (2001, 227–8). Barry Dainton notes that, "Coordinate independence... is a criterion of a quantity being physically real as opposed to an artifact of a particular mode of representation" (2010, 225). David Baker says that "only... invariants [under symmetry transformations] are physically real" (2010, 1161). Travis Norsen says that since simultaneity is frame-dependent in special relativity, according to this theory any claim involving the instantaneous configuration of particles "is literally meaningless" (2017, 12).

These statements are fine as far as they go, but it is worth being a bit more careful. Frame-dependent quantities, like velocity in Newtonian physics or simultaneity in special relativity, are not well-defined, meaningful, or physically real *independent of* the specification of a reference frame or coordinate system; they needn't thereby be wholly meaningless or unreal. As an analogy, we cannot ask what are "the" x- or y- coordinates of a given point in the Euclidean plane: this is not a meaningful question, since the answer depends on the coordinate system, many choices of which are equally legitimate. But we can meaningfully ask what are the point's coordinates in a particular coordinate system: once we specify the coordinate system, there will be a fact about this. Similarly, it does not make sense to ask whether an object is "really moving" in Newtonian mechanics or whether two spacelike separated events are "really simultaneous" in special relativity: these are not meaningful questions, since the answer depends on the reference frame, many choices of which are equally legitimate. But we *can* meaningfully ask whether a given object is moving or two events are simultaneous according to a particular observer or reference frame: there will be a real physical fact about these things. We might put this by saying that *absolute* velocity in Newtonian physics and *absolute* simultaneity in special relativity are not physically real; but this is just to say that velocity and simultaneity full stop, without the specification of a frame, are not physically real.

As a more general point, it is too quick to say that just because a feature or quantity depends to some extent on an arbitrary choice in description, it is therefore entirely meaningless or unreal, wholly detached from physical reality. The choice of unit of length is conventional or arbitrary; but once given a choice, an object's length is not arbitrary or meaningless: the object will have a particular length relative to that choice of unit. Velocity in Newtonian mechanics is frame-dependent; but once given a choice of frame, an object will be moving in a certain way with respect to it: there is a physical fact about this. Such quantities depend on, they are in part about, our conventional descriptions of the world. But they *also* depend on, and are *also* in part about, the world itself. Recall Lange's statement that, "frame-dependent quantities reflect *not only* the world, but *also* the chosen reference frame" (2017, 142; my emphasis). As a result, frame- or coordinate-dependent features can tell us about the world, so long as we are careful. (We might want to say that description-dependent facts are real but not fundamental: they hold in virtue of the description-independent facts. This will depend on one's view on fundamentality and on the relationship between the fundamental and the nonfundamental, things I aim to remain neutral about as much as possible.)

This may sound like a minor point, and one with which the authors quoted above will surely agree. There is a reason for emphasizing it, which will become clearer in later chapters. The reason has to do with the following. The idea that frame- or coordinate-dependence renders a quantity or feature meaningless or physically unreal can lead to the further thought that the mere mention of reference frames or coordinate systems implies that we are not characterizing physical reality, but only our own description of or perspective on reality, and likewise that any phenomenon described from the perspective of a reference frame or coordinate system is wholly unreal or unphysical, merely a feature of our own representation or perspective. However, this thought is not right, and it can lead to incorrect conclusions about the physics.[22]

Consider John S. Bell's discussion in "How to Teach Special Relativity" (1987a). Bell notes that there must be a physical explanation of the phenomenon of

[22] Ismael makes a related point in discussing the nature of time in physics versus experience, noting that, "It's difficult to say how 'perspectival' came to be associated with 'unreal'," an association that "has been one of the most insidious and confusing aspects of the physical discussion of time" (2016, 119).

length contraction, of the fact that observers in relatively moving reference frames will measure different lengths of objects (in particular, that a relatively moving observer will measure a shorter length, in the direction of the motion, than an observer at rest with respect to the object)—an explanation in terms of systems' atomic constituents and interatomic forces. The fact that this phenomenon is "relative to a frame" does not make it entirely physically unreal or merely perspectival, just an artifact of description. Indeed, it was the mistaken thought that this kind of thing is frame-dependent and *therefore* merely perspectival or physically unreal that led Bell's colleagues to deny that the string between the accelerating rockets, in his example, would break. As Bell points out, there must be a physical account of the string's breaking, and this is compatible with the fact that length contraction is relative to or dependent on the reference frame.

Now, in some cases, we do conclude that frame- or otherwise description-dependent features or quantities or phenomena are not physically real. (Consider pseudo forces in Newtonian mechanics or gravitational forces in general relativity.[23]) The point remains that it is not the description-dependence *per se* that yields the conclusion, as we can see from Bell's example of length contraction in special relativity. Or consider the potentials in classical electromagnetism, which "are only mathematical conveniences, and arbitrary to a high degree, made definite only by the imposition of one convention or another" (Bell, 2004, 234). We deny the physical reality of the potentials in classical electromagnetism because different potentials-based descriptions appear equally capable of capturing the physical facts; so that any choice we do make seems a conventional choice in description, not corresponding to anything in physical reality. Nonetheless, it is possible to interpret the potentials as physically real (which some people think we should do in the quantum case), and according to such an interpretation, this won't be an arbitrary choice among equally good descriptions. What we regard as a conventional choice in description will depend on what we take to be physically real, in other words, but likewise what we take to be physically real will depend on what we regard as a conventional choice in description. The way to break into the circle is by means of some initial physical posits, which

[23] Or consider what Maudlin calls "coordinate-based Lorentz-Fitzgerald contraction," a coordinate-dependent effect, akin to an abstract velocity boost that mathematically switches reference frames rather than physically altering any system's velocity, which is "not, in any straightforward sense, the physical contraction of anything... nothing is subjected to any forces and nothing 'shrinks'" (2012, 99). This is as opposed to the "physical Lorentz-Fitzgerald contraction" in Bell's example. The former phenomenon is merely a fact about coordinate systems and how they relate to one another mathematically—it is "an abstract descriptive change, not a physical change" due to real physical forces (Maudlin, 2012, 114). I would add that even in the former case, there will be a physical account of how objects' lengths differ according to different reference frames, for the transformation equations expressing how the coordinates of different reference frames relate to one another mathematically flow from the underlying spacetime structure; in that sense, even this kind of thing is not wholly detached from physical reality.

will help to distinguish what is mere convention from what is not. In the case of classical electromagnetism, we assume—not without good scientific reason—that the theory is fundamentally about the fields, not the potentials. (More on this in Chapter 7.) This is why I said that such things *needn't* be completely unreal, that this doesn't follow from the description-dependence *alone*.

The idea of structure is also similar to that of intrinsic properties, although the extent of the similarity will depend on how one explicates the notion of an intrinsic property, which I won't take a stand on here. Intuitively, an intrinsic property is a feature that an object has in itself, regardless of anything else. David Lewis, in discussing intrinsic properties, says things that are reminiscent of what I have said about structure:

> A sentence or statement or proposition that ascribes intrinsic properties to something is entirely about that thing; whereas an ascription of extrinsic properties to something is not entirely about that thing.... A thing has its intrinsic properties in virtue of the way that thing itself, and nothing else, is. Not so for extrinsic properties. (Lewis, 1983, 197)

There are difficulties with providing an analysis of intrinsic properties that make it hard to say whether this amounts to the idea of structure; similarly, whether all and only the intrinsic properties of a thing get at its structure. Structure concerns features that are independent of arbitrary choices in description in particular. Although these kinds of features often do seem to count as intrinsic, there may be intrinsic features that are not structural in my sense. (Consider "object o has velocity v relative to inertial reference frame S in Newtonian world w." Intuitively this is not a statement about structure. Does it ascribe an intrinsic feature to o? Hard to say. It is not wholly about the object itself, for it is in part about the descriptive apparatus we bring to bear. At the same time, an object all alone in a world can arguably have the feature (setting aside concerns familiar from the traditional debate over relationalism about motion).) Suffice it to say that there is some connection between intrinsic properties and structure, but the extent of the connection will depend on one's preferred account of intrinsic properties, which is not something I address here.

A final reminder on two points. First (as mentioned in Chapter 1), there is a difference between mathematical objects or structure, and physical objects or structure. To say that a physical object possesses a certain structure is to say that it has features that are well-represented by the relevant mathematical structure. It does not mean either hypostatizing mathematical objects or saying that physical entities are mathematical things. Second (as mentioned at the beginning of this chapter), this notion of structure is not novel, but is implicit in much of our theorizing about physics and mathematics, as evidenced by the examples here.

2.4 Comparing structures

In both mathematics and physics, then, we talk of the structure of different kinds of things—spaces, objects, worlds, even, we will see, entire physical theories. What is more, we also distinguish among different types of structure, which can be organized into a hierarchy of different "levels" of structure. This, in turn, allows us to compare different things (spaces, objects, worlds, theories) with respect to their relative strengths or amounts of structure.

I will illustrate this here by describing a few different types of mathematical structure and how they are related to one another. This is not intended to be an exhaustive examination of all the kinds of structure there are and all their interrelations. Nor will the resulting comparisons of structure be completely clear-cut in every single case. I aim to show that this means of comparison is nonetheless familiar, intuitive, and useful, indeed for the most part clear-cut. As we will see in the following chapter, we put these structural comparisons to good use in a variety of inferences we make about physics.

Start with the most basic kind of structure, a set of points. (The elements of a set do not have to be points, and the bottom-level structure does not have to be a set structure, but this is most familiar and I will assume it here.[24]) A set is a bare collection of elements with no further structure. This lowest level of structure consists of facts about cardinality and the membership relation, and other set-theoretic notions that can be defined in terms of these (subset, union, intersection, and so on). No further mathematical objects or notions are defined at this level of structure. As one author puts it, "by itself, a set has no structure other than what it contains" (Isham, 2003, 59), "its only general mathematical property being the cardinal number" (Isham, 1994, 10).

We can then add structure to a set by specifying further primitive notions and defining additional mathematical concepts in terms of them. Doing so takes us higher up in the hierarchy. For example, given a set of points, we can define a topology on that set by specifying the open subsets, which are subject to certain axioms. This endows the set with further structure, a topological structure, and turns the set into a topological space. A topological space is a set in which a certain family of subsets is distinguished as the open sets: "A topological space X is a set-with-structure, where in this case the 'structure' consists of a specified collection of the subsets of X, namely the collection of all open sets" (Mac Lane, 1986, 33).

A topological space is in essence a special kind of set, one in which the notion of an open set is recognized; one in which the subsets are specified, as is the case for any set, but in which there is also specified a special collection of subsets, the open sets. This special collection of subsets can then be used to define various

[24] Arntzenius (2012, Ch. 4) discusses the idea of a "gunky," or pointless, structure at the bottom level, while giving reasons to be skeptical of its viability, especially for physics.

topological notions such as the continuity of curves and the neighborhoods or nearness of points.[25] A bare set lacks the notions of open and closed subsets supplied by a topology. A topology in this way *adds* structure to a set, yielding facts about open and closed sets, as well as continuity, connectedness, and convergence. Absent a topology, there simply are no facts about whether a given subset of points forms a continuous curve, for instance: the notion of a "continuous set of points" is not recognized or meaningful without this level of structure.

Intuitively, a topological space has more structure than its underlying set, and we can say more generally that a topology is more structure than a set, that a topological space has more structure than a bare set. In defining a topology, we add structure to a set, in that new mathematical objects or concepts are specified or defined (open sets, the neighborhoods of points, and so on); new facts hold and further distinctions are made (whether or not a given set of points forms a continuous curve, whether a subset is open or closed, and so on); new notions make sense that do not make sense for a bare set (continuous function, the convergence of a sequence, and so on). A topological space is in this way a special case or a special kind of set, one in which more notions are specified, more facts are countenanced, more distinctions are recognized.

Another way to see the relationship between these two types of structure is to note that every topological space has an underlying set structure, but not every set has a topological structure. We can define a set without mentioning any topological notions, whereas a topology effectively assumes or presupposes a set structure: a topology specifies, from among all the subsets specified by the set structure, which are the open sets. A bare set does not presuppose or require, let alone recognize, facts about openness and continuity and so on. A set structure is in this way conceptually prior to a topology. Another way to see the relationship: a topology must satisfy additional constraints beyond those of a set, namely the axioms that must be satisfied by certain of the subsets. In all these ways, a topology is intuitively *more structure* than a set. A topology is a stronger structure; it lies at a higher level of structure.

We are starting to see that levels higher up in the hierarchy contain additional structure in that further mathematical objects or notions are recognized, defined, or meaningful; additional mathematical facts hold; further distinctions are drawn. Since the higher-level structures retain the notions specified at lower levels while adding new ones, we can also regard a given higher-level structure as a "special case" or a "special type" of lower-level structure. We will see other ways of conceptualizing the relationship between the different levels of structure in a

[25] Standard topology often proceeds in this way, with "open set" as the primitive notion, although there are standard topologies that take either the notion of "closed set" or "neighborhood" instead to be fundamental, defining the other notions in terms of it. See Maudlin (2014a) for a non-standard topology, which does not take any of these notions as primitive.

moment. (The "levels" idea is intended to be helpful and intuitive, if informal. This way of speaking does appear in many places. But it needn't be taken too literally in order to grasp the basic idea of different types of structure that can be organized into a kind of hierarchy by means of their relative amounts. Examples are in the footnote of sources in philosophy, physics, and mathematics that intimate the idea, some of which explicitly mention "levels of structure."[26])

A topological structure is too minimal a structure on which to do calculus, which is central to physics and its differential equations of motion. For that we need further structure, a differentiable structure, which allows us to coordinatize the space and define the differentiation of functions on it. To add this structure, first of all assume a particular kind of topological space called a topological manifold. A topological manifold is a topological space that has the structure of a manifold: it is everywhere locally "like" $\mathbb{R}^n$; that is, every point has a neighborhood that is topologically the same as, or homeomorphic to, an open set in $\mathbb{R}^n$.[27] (Not all topological spaces are topological manifolds.) This is the kind of space on which we can introduce a local coordinate chart, or coordinate patch, at each point, allowing us to label each point uniquely by means of n numbers. If we then smoothly stitch these local charts together into a smooth "atlas," for which the overlap maps or transition functions between overlapping coordinate charts are smooth, it becomes a differentiable manifold. A differentiable, also called smooth,

[26] Sklar (1974, 48–54) discusses several "levels of abstraction" away from a given geometry, obtained by ignoring various notions in turn, a process that leads us through different "level[s] of structure"; as we remove features, "we will have described a structure of which, in general, the original 'full' structure will be only one particular example" (1974, 49). Friedman says that, "given any particular geometrical space or manifold, we can distinguish various levels of structure" (1983, 10). (Note that what Friedman calls "higher" versus "lower" levels is the reverse of mine.) Mac Lane discusses how "many Mathematical notions can be described as set-with-structure" (1986, 34), a basic set with additional structures defined on it. Earman (1989, Ch. 2) describes how we can build up different spacetime structures by adding further structures to a manifold, and compare the structures that result via the quantities that are defined in them. Stachel (1993, 135) explains how we can perform a "sequence of abstractions" (on a model of a theory) by removing different types or levels of structure. Isham (1994, 10–11) discusses a hierarchy of spacetime structures and the idea of different amounts of spacetime structure. Lee describes different "layers of structure" (2003, 2). Malament (2012, 132) mentions the idea of "a spacetime model (M, g_{ab}) exhibiting several levels of geometric structure." Maudlin (2012, Ch. 1) discusses "different sorts of geometrical structure, which form a hierarchy" or "levels of structure" (2012, 5; 7). Some books in mathematical physics that suggest the idea are Schutz (1980) (who describes various "level[s] of geometry" (23)); Geroch (1985); Isham (2003). (A different hierarchy, organized according to different criteria, is in Curiel (2017, 91).) The general idea of adding structure to an existing one in order to obtain a special case or stronger structure, one in which more mathematical notions or concepts are defined or make sense, is ubiquitous. To give one example, after defining a vector space, one book adds that, "In many vector spaces there are additional operations such as taking an inner (dot) product, but this is extra structure over and above the elementary concept of a vector space"; similarly, a tangent bundle is "a specific example of a 'fiber bundle', which is endowed with some extra mathematical structure" (Carroll, 2004, 16). Many more examples could be given.

[27] A homeomorphism is mapping that preserves topological properties (a continuous one-to-one mapping with a continuous inverse), so that spaces that are homeomorphic are the same from the point of view of topology, possessing the same topological structure. A topological manifold will have separability and countability conditions that allow it to be coordinatized in this way (it is required to be Hausdorff and usually also paracompact or second countable).

manifold is a space that is coordinatized in this way, that is equipped with an atlas, giving it a global differentiable structure that allows us to define derivatives and other central notions from calculus. We can then regard a differentiable manifold as "a set with two layers of structure: first a topology, then a smooth structure" (Lee, 2003, 2).[28]

Intuitively, a differentiable structure is more structure than a topology. More mathematical concepts are defined; more mathematical facts hold; further mathematical distinctions are drawn. A smooth manifold "is a topological space with some additional structures" (Stachel, 1993, 131), "a topological manifold with some extra structure in addition to its topology, which will allow us to decide which functions on the manifold are smooth" (Lee, 2003, 11). A topology specifies which curves or functions are continuous. A differentiable structure further distinguishes, from among the curves that are continuous according to the topology, those that have sharp bends from those that are smooth, and to what degree they are smooth (how many times differentiable); similarly, which functions on a space are differentiable, and to what degree. A topology alone will not do this. A topological structure does not countenance facts about the smoothness of curves or the differentiability of functions. Such notions are simply not recognized or defined at that level of structure; whereas in a differentiable manifold M, "there is a meaningful notion of *differentiability* for functions defined on M (unlike a simple topological space, which has a notion of continuity but not of differentiability)" (Friedman, 1983, 340).

(It is clear that no topological property alone will be capable of yielding a suitable notion of smoothness (Lee, 2003, Ch. 1). Topology concerns the features that are invariant under homeomorphisms (continuous one–one mappings with a continuous inverse), and no plausible notion of smoothness will be so invariant. Consider a circle and a square in the plane. These are homeomorphic: they can be "smoothly deformed" into one another. (Smoothly: stretching, squeezing, and shearing are allowed, but no tearing or pasting.) Yet the circle is smooth whereas the square is not; the square has corners, the circle does not. More generally, derivatives of functions won't be invariant under homeomorphisms. In other words, the kind of structure that we need to do calculus, which will distinguish between a curve or function with sharp corners and one without, which will allow us to define the differentiation and integration of functions, must be structure over and above a topology.)

[28] "Smooth" or "differentiable" here generally means C^∞ (infinitely times differentiable) unless otherwise indicated. The above assumes the standard definition of a differentiable manifold. Arntzenius and Dorr (2012) define this structure without any mention of coordinates or charts, in particular without having to specify the topology by way of the topology of the real numbers. We might say that a differentiable manifold is either a space that has been coordinatized in the above way, or one that (has the structure that) allows us to do so.

Another way to see the relationship between these two types of structure is to note that every differentiable manifold has an underlying topology, but not every topological space has a differentiable structure defined on it. There are moreover topological manifolds that can be given different (non-diffeomorphic[29]) smooth structures (such as the so-called "exotic $\mathbb{R}^4$"s (Freedman and Taylor, 1986)), as well as topological manifolds that do not admit any (global) smooth structure at all. In other words, a topological structure does not determine or presuppose a differentiable structure, whereas a differentiable structure does presuppose or determine a topological structure. A differentiable manifold is in this way a special case or "a special type of topological space" (Isham, 2003, 2). Note for future reference that a differentiable structure is intuitively additional structure over and above a topology, even though not every topological space can be given a differentiable structure: only the topological manifolds can be given this further structure.

(If we consider the mathematical structures lower down to be more fundamental than those higher up, then the mathematical idea of relative fundamentality is interestingly different from a familiar philosophical one. According to a familiar metaphysical conception of relative fundamentality, things that are more fundamental in some sense necessitate things that are less fundamental. For the mathematical hierarchy, by contrast, things at higher levels constrain things lower down.[30] I refrain from referring to these levels of structure in terms of relative fundamentality to avoid any confusion on that front.)

Given a differentiable manifold, we can go on to define further types of structure. We can define an affine structure, for instance, which provides a standard of "straightness" of curves: this structure distinguishes, from among the smooth curves specified by the differentiable structure, the ones that are straight (the geodesics) from those that are not. An affine structure is a differentiable structure with the added requirement that the charts have a straight-line structure and the transition functions between overlapping charts are affine transformations, which preserve that structure, preserving the straightness and parallelism of lines. Equivalently, an affine manifold is a differentiable manifold with the addition of an affine connection, allowing us to define the parallel transport of vectors (the straight lines being those along which vectors can be parallel transported).

Intuitively, an affine structure is over and above the structure of a differentiable manifold. More mathematical notions are defined (straight versus curved line, the parallelism of lines), more mathematical facts or distinctions are countenanced (whether or not a curve is a geodesic, whether a vector is parallel transported),

[29] A diffeomorphism is a mapping that preserves differentiable structure, a differentiable map with a differentiable (to the same degree) inverse. The above means that certain topological manifolds can be given distinct, inequivalent smooth structures.

[30] Compare Dorr (2011, 144–5).

facts and concepts and distinctions not given by a differentiable structure alone. In the words of one book, "On a simple differentiable manifold the question of parallelism at different points does not even make sense, since there are no 'markers' or rules for moving vectors around in a parallel manner. One must add more structure—called an 'affine connection'—to the manifold in order to define an absolute parallelism" (Schutz, 1980, 76).

We could then go on to add a metric, which "creates even more structure than the affine connection" (Schutz, 1980, 201). A metric is a function that gives distances between points, assigning a real number to each pair of points in such a way as to satisfy the constraints of a distance measure. The metric says, given the geodesics specified by the affine structure, what is the distance between any two points measured along such a curve between them (on the usual conception of distance in differential geometry and physics: below). The straightness of curves, by contrast, can be specified without a notion of distance. In this way a metric presupposes an affine structure, but not vice versa. A metric adds structure to an affine space: more mathematical notions are defined (distances between points, lengths along curves), more mathematical facts are countenanced and distinctions are drawn. (There is a distinct concept of a geodesic as a local distance-minimizing path, the "shortest" rather than "straightest" path between two points, which may seem to suggest that an affine structure presupposes a metric structure rather than the other way around; however, it is possible to define an affine structure without assuming any metrical notions, so that the metric is the higher level of structure.[31]) At the same time, both an affine and a metric structure are over and above a differentiable structure, in that "the notions of distance between points and straight lines (or shortest paths) are not part of the idea of a manifold but arise as a consequence of additional structure, which may or may not be assumed and in any case is not unique" (Bishop and Goldberg, 1980, 19).

We see that one way of capturing the relationship among the different types or levels of structure is to say that as we go up the hierarchy of structures, more mathematical notions are defined and facts are countenanced, more mathematical distinctions are drawn. Another is to say that structures higher up determine or induce structures lower down, but not vice versa. An example of this is given by the relationship between a metric and a topology. A topology is definable

[31] Compare Friedman (1983, 349): "geodesics can be regarded either as 'shortest' curves or as 'straightest' curves. We can give a precise definition of 'shortest' curve by introducing a *metric* on [manifold] *M*, but there is a conceptually distinct notion of 'straightest' curve that can be introduced independently of a metric." As Sklar puts it, in an affine space, given a curve *C* and two points on it, *P* and *Q*, "we simply cannot ask, in general, 'How far is it along *C* from *P* to *Q*?' We can however ask, 'Is *C* the straightest curve between *P* and *Q*?' and expect to get an answer" (1974, 50). In a metric space, by contrast, we can expect an answer to the first question. Sklar notes that in an affine space there is a limited notion of distance in the sense of ratios of intervals along geodesics, but there is no general notion of distance along a curve between any two points, which is "simply not defined in such a space" (1974, 50).

independently of a metric: a topology specifies the open subsets and the neighborhoods of points without requiring a notion of distance. A metric space has a notion of distance defined on it, which can then be used to define a topology in the natural way (by means of the open balls, sets of points that are within a distance *d* from a given point).[32] On its own, a topological space does not have a notion of distance, and for any given (metrizable) topological space, different distance functions, or none at all, could be defined: the topology does not determine a metric, whereas the metric does determine or induce a topology. There are furthermore (non-metrizable) topological spaces that cannot be given a metric structure. Not every topological space is a metric space, in other words, but every metric space is effectively a topological space. (Another way to put this is that not all topologies can be generated by a metric, only the metrizable ones can be.) All of which is to say that "a metric space is a special case of a general topological space" (Isham, 2003, 14); that, "Topology is a more 'primitive' concept than distance" (Schutz, 1980, 5); that a metric is structure over and above a topology. Here is one mathematician expressing the overall idea:

> The fact that different operations can [sometimes] be defined over the same underlying set creates a *hierarchy of structures*. This means that certain notions of space are more refined than others, or, to put it differently, spaces of certain sorts are automatically also spaces in any of the underlying coarser notions. Thus, for example, . . . a smooth manifold is also a topological space, or a metric space is also a topological space. Usually none of these implications can be reversed; namely, there are topological spaces that cannot be made into metric spaces and topological spaces that are not smooth manifolds. (Marcolli, 2020, 43)

Hence there is another way of characterizing the relationship among the different types of structure, which has been implicit so far: mathematical objects or concepts defined at levels higher up presuppose or assume ones specified at levels lower down. A metric presupposes a topology in that a metric gives distances

[32] That is, "Every metric or pseudo-metric space is a topological space since the balls $B_\epsilon := \{y \in X | d(x, y) < \epsilon\}$ are open ... and form a basis for the neighborhoods of *x*" (Isham, 2003, 33), where *d* is the distance function. By contrast, we can define a topological notion of nearness without a metric function being defined (see Isham, 2003, Sec. 1.4 for one way of doing so). In various spaces there can be a topology that differs from the one induced by means of the open balls of the metric (which can even be the physically significant topology; perhaps this is what Carroll means when he says that, "the metric we use in general relativity cannot be used to define a topology" (2004, 71)). Nonetheless, it remains the case that any metric induces or gives rise to a topology, that any metric induces a natural topology in this way, and not vice versa. (There are also spaces, including certain spacetimes of general relativity, for which the topology induced by the metric may not be the same as the underlying manifold topology; however, it has been shown that if a relativistic spacetime is strongly causal, then we can recover the manifold topology with something like the open balls of the metric (Beem et al., 1996, 144, Theorem 4.9). Thanks to Gordon Belot for discussion and the reference.)

along continuous curves, and a topology is needed to specify which are the continuous curves. Similarly, a metric gives distances along curves by adding up the lengths of segments between nearby points, and without a topology there is no sense of the nearness, or neighborhoods, of points. A metric is a higher level of structure in that it presupposes the lower-level topological notions of continuity and neighborhoods; a topology, however, does not presuppose metrical notions. This may seem to run counter to the idea of a higher-level structure's inducing a lower-level one: a metric induces a topology, which might make it sound as though the (induced) topology in turn presupposes the metric. But in fact it amounts to the same idea, namely, that once we have defined a metric, there is thereby already implicitly a topology. A metric presupposes a topology, as evidenced by the fact that once we define a metric, a natural topology is thereby induced, yet it is not the case that once we have a topology, a metric is thereby induced; that every metric space has an underlying topology, but not every topology has a metric defined on it; that every metric induces a topology, but not vice versa.[33]

Note that the conception of distance in play here is the one that is used in standard differential geometry and is most relevant to and familiar from physics. It is what Phillip Bricker (1993) calls the "Gaussian" as opposed to the "intrinsic" conception. According to the intrinsic conception, the distance between two points depends solely on the properties of the points: it is a feature of the two points alone, intrinsic to them, regardless of anything having to do with the space in which they are embedded, including anything having to do with the topology of the space. Distance relations are primitive, in other words, and other features of the space are defined in terms of them. According to the Gaussian conception, by contrast, the distance between two points is given in terms of the length of a continuous path (the shortest path; or in Lorentz signature, the longest) between them. The lengths of paths is the more basic notion, and the distance between two points is not intrinsic to the pair of points alone, but depends on the nature of the surrounding space. (That said, even on the intrinsic conception of distance the topology can be seen as conceptually more basic in various senses: every metric determines or induces a topology but not vice versa; different metrics can induce the same topology; not all topologies can be generated by a metric.) As a more general note, the hierarchy of structures I am discussing stems from the usual way of developing and defining these notions in differential geometry and mathematical physics. This is not to deny that there can be other ways of developing things, but this is a natural and familiar way, especially when it comes to physics; it is the one that lies behind the familiar inferences discussed in Chapter 3.

[33] You may worry that the topology induced by the metric need not be the unique topology one can define (cf. note 32). As an example, we can put the discrete topology on the reals even while using $|x - y|$ as the distance function. That said, for a given metric there is a unique topology that is the coarsest topology with respect to which the metric is continuous, which is in that sense the natural topology. Thanks to Laura Ruetsche for the example and the response.

One final way of seeing the hierarchical form of these structures is by considering the transformations or mappings that preserve a given structure, that is, the isomorphisms of the structure, also called the automorphisms, the isomorphisms from the structure onto itself. We saw that we can characterize the structure of the Euclidean plane by means of the quantities or features that are invariant under allowable coordinate transformations. It turns out that we can characterize other structures similarly, by means of the invariant quantities under the relevant structure-preserving transformations. This is in the spirit of Klein's (1892) Erlangen program for geometry. Klein suggested that any geometry can be identified by means of the transformations that preserve the structure, likewise by the quantities that are invariant under the group of those transformations—as Euclidean geometry concerns those features that are invariant under rigid translations, rotations, and reflections.[34]

Given this idea, the hierarchy of structures should correspond to a similar hierarchy of (groups of) transformations. Intuitively, a wider group of structure-preserving transformations means that there are fewer features or quantities to be preserved, so that a wider transformation group is indicative of a lesser or lower-level structure. Comparing the sizes of the groups of structure-preserving transformations then yields a measure of relative amounts of structure.[35]

A set structure, for instance, is invariant under bijections, one-to-one and onto mappings that preserve cardinality, so that we can identify a set structure as comprising those features that are invariant under bijections. When it comes to topological structure, further constraints must be met by a mapping in order for it to count as structure-preserving: additional features must be preserved. The relevant structure-preserving mappings are in this case the homeomorphisms, continuous bijections with a continuous inverse, which map open sets to open sets and continuous curves to continuous curves. We can then identify a topological structure as comprising those features that are invariant under homeomorphisms. (Hence the idea of topology as "rubber-sheet geometry": topological features are unaltered by continuous transformations, which include stretching, squeezing, and shearing.)

A comparison of the transformations that preserve these two types of structure yields the same verdict as above on their relative amounts of structure.

[34] The Kleinian conception of geometry does not encompass every kind of structure. We will see an example in Chapter 4.

[35] The group of structure-preserving transformations of a higher-level structure may in general form a subgroup of the group of transformations preserving a lower-level structure: see Wilhelm (2021) on this way of comparing structures. Wallace (2019) defends this conception of structure and mentions this way of measuring relative amounts of structure. He says: "As the group [of structure-preserving transformations] is made larger, the space becomes less structured" (2019, 127); for example, within a hierarchy of pre-relativistic spacetimes, "Each [structure-defining] group is a subgroup of those below it, so that we can see the move from one spacetime to the next as a successive discarding of structure" (2019, 128).

A homeomorphism, the kind of mapping that preserves topological structure, will not alter the underlying set structure, whereas a bijection need not preserve the topological structure. Every homeomorphism is a bijection, in other words, but not every bijection is a homeomorphism. The bijections form a wider group of transformations than the homeomorphisms; they preserve a lower level of structure.

Or consider a comparison of differentiable and topological structure. A differentiable structure is invariant under diffeomorphisms, which map smooth curves to smooth curves, so that a differentiable structure comprises the features that are invariant under the group of diffeomorphisms. In order to be a diffeomorphism, a function and its inverse must be differentiable (to the same degree). In order to be a homeomorphism, the function and its inverse must only be continuous. A comparison of structure-preserving mappings then yields the same verdict as above on their relative amounts of structure. Every diffeomorphism is a homeomorphism, but not every homeomorphism is a diffeomorphism. A topological transformation can alter the differentiable structure, as a square can be continuously transformed into a circle despite their different differentiable structures. The diffeomorphisms form a narrower group than the homeomorphisms; being preserved by a diffeomorphism is a stronger condition than being preserved by a homeomorphism. A diffeomorphism preserves a higher level of structure.

For similar reasons, an affine structure lies at a higher level of structure than both a topological and a differentiable structure, but at a lower level than a metric structure. An affine structure is invariant under affine transformations, which map straight lines to straight lines. An affine transformation preserves the topological features of a space, but a topological transformation that preserves facts about continuity can alter the straightness of lines. Metric features are invariant under isometries, rigid transformations that preserve distances. Although every isometry is an affine transformation, not every affine transformation is an isometry. Stretching or squeezing a space preserves the straightness of lines, but alters the lengths of paths and the distances between points. Isometries preserve a higher level of structure than that preserved by affine transformations. (A uniform stretching or squeezing is an affine transformation but not an isometry, preserving the straightness of lines but not distances. Whether we take such a mapping to count as preserving the metric structure will depend on whether we identify "the" metric structure with the distances given by a metric function (which are altered by such a transformation) or by the scale-invariant ratios between distances (which are not altered by such a transformation). On the latter idea—which will be preferable to many philosophers on the grounds that it does not single out a privileged scale or unit of distance—there will be a family of metric tensors or functions (and corresponding isometries), all of which represent the underlying metrical facts equally well.)

Notice how the above suggests that we cannot say that two objects or spaces are isomorphic, full stop. We can only say that they are isomorphic *with respect to* a given type of structure, when there is a mapping between them that preserves a certain type of structure, in which case we may say they are equivalent in terms of that structure. This further suggests that two objects or spaces can be isomorphic or equivalent with respect to a certain type of structure, while at the same time being non-isomorphic or inequivalent with respect to another type of structure. We will see this illustrated later on. (We could put this point, indeed much of the discussion in this section, in terms of category theory, where the kind of mapping that preserves a given structure is defined relative to a category. Recent work in philosophy of physics has advocated using category theory as the proper framework for comparing structures. I set this aside since the categorical approach requires technical machinery unnecessary for the basic points I wish to make here.)

On any of these ways of comparing of structure, a symmetry will generally indicate less structure, as a Euclidean plane with a preferred location or direction (an orientation) has more structure than an otherwise-similar plane without. The preferred location or direction is a further piece of structure that can be added to the symmetric plane. This verdict comes from comparing the groups of structure-preserving transformations as well. As one author says, "The plane with an orientation has more structure—namely, the choice of the orientation. At the same time, it has less symmetry; the automorphism group of the oriented plane is the group of all proper rigid motions (i.e., no reflections), while that of the unoriented plane is the group of all rigid motions, including the reflections" (Mac Lane, 1986, 84).

There are then several, not wholly distinct, ways of conceptualizing the relationship between different types or levels of structure, which usually (note 34) converge on a verdict as to their relative amounts of structure. As we go up the hierarchy, we gather additional structure, in a few senses. (1) More mathematical notions make sense, are defined or meaningful or recognized; more mathematical facts hold; more distinctions are countenanced or drawn. (2) Higher-level concepts or notions do not make sense or cannot be defined absent various lower-level notions, whereas lower-level notions can be specified or defined independently of higher-level ones. (3) Types of structure higher up assume or presuppose or require structures lower down. Types of structure higher up are not similarly assumed or presupposed or required by structures lower down. (4) Structures higher up constrain or induce structures lower down; lower-level structures do not similarly constrain or induce higher-level ones. (5) A higher-level structure is less general, a special case of a lower-level structure, satisfying further conditions. (6) The associated group of structure-preserving transformations becomes narrower.

In all, there are different types of structure, and there is an intuitive hierarchy of structures, organized according to their relative strengths or amounts. We can often add structure to a mathematical space or object to get a different kind of object, which is higher up in the hierarchy, possessing a stronger structure; one that is a special case of the original, in which more mathematical notions are countenanced or defined. (Compare Earman on different spacetime structures in particular: "As the space-time structure becomes richer, the symmetries become narrower, the list of absolute quantities increases, and more and more questions about motion become meaningful" (1989, 36).)

These methods of comparison will not always yield a verdict that one structure is stronger than another. As an example, consider affine structure and conformal (angle) structure. A conformal, or angle-preserving, transformation needn't preserve the straightness of lines: not all conformal transformations are affine transformations, which seems to suggest that affine structure is a stronger, higher-level structure compared to conformal structure (as a metric structure, preserved under isometries, is a higher-level structure compared to an affine structure, preserved under affine transformations, evidenced by the fact that not all affine transformations are isometries). However, an affine transformation, in turn, needn't preserve angles: not all affine transformations are conformal transformations. In fact, neither affine nor conformal structure is stronger than the other, though it is unclear whether we should conclude that they have the same amount of structure or that these structures are rather incomparable. The hierarchy in Friedman (1983, 12), for one, depicts them in such a way that they appear to be incomparable (at the same time, each of these structures is presupposed by, and so amounts to less structure than, a metric structure).[36]

Not only will there be cases for which we cannot say that one structure is stronger than another, nor even perhaps that they are equally strong, there may be cases for which these ways of measuring structure and comparing different structures are not absolutely precise or clear-cut. An example: take a mere topological space, but now add one distinguished point. And compare that with a metric space. One might want to say that the latter space has more structure, in a sense closely related to what I have been discussing, even though the set of notions that make sense in this space is not a proper subset of the set of notions that make sense in the other.[37] (Another example will come up in Chapter 3.) None of this means these comparisons are without value. The examples discussed above, with more to come in later chapters, reveal that we do often compare things in these ways in mathematics and physics. These comparisons of structure are intuitive and familiar, if sometimes inexplicit; they work well in a variety of cases; they are for the most part clear-cut. These comparisons form the basis for some general principles of physical theorizing, discussed in Chapter 3.

[36] Further examples of potentially incomparable structures are discussed in Swanson and Halvorson (2012); Curiel (2014); Barrett (2015a,b); Wilhelm (2021).

[37] Thanks to Ted Sider for the example. Analogously, a symplectic structure is intuitively less structure than a metric (as I argue in North (2009)), even though the former possesses an orientation. (See Swanson and Halvorson (2012); Barrett (2020a) for argument that these structural comparisons do need to be made more precise.)

3
Inferences about Structure

> [R]ules and particular inferences alike are justified by being brought into agreement with each other . . . ; and in the agreement achieved lies the only justification needed for either.
>
> Nelson Goodman (1955, 64)

3.1 Inference rules

Chapter 2 outlined the notion of structure I have in mind and pointed to ways in which this notion is implicit in our thinking in physics and mathematics. Learning about the structure of a thing—an object, space, world—tells us about the intrinsic nature of that thing, what it in itself is like, apart from our descriptions or representations of it.

In this chapter, I look in more detail at some familiar inferences about physics that make use of this idea and the related one of comparing different amounts of structure. This will reveal a few rules or principles we commonly rely on when making these inferences. It is the aim of this chapter to bring these principles, often tacitly assumed, to the fore.

The inferences I discuss will demonstrate that we do familiarly and successfully rely on these rules, and that we are reasonable in doing so. These are principles we take ourselves to be justified in adhering to, not because we have a general argument that they are bound to yield the right results, but because they tend to yield conclusions we generally accept as reasonable. As Nelson Goodman puts it, by "making mutual adjustments" (1955, 64) between the general rules and the particular inferences they yield, we bring the rules and inferences together into what has come to be called a "reflective equilibrium," and in so doing we find a justification for both. This is a way of justifying the rules of inference, where the particular inferences, in turn, are justified by the fact that they conform to these general rules.

There are questions that can be raised about the reasonableness of this method of justification, which I won't address here. For instance, this method will be unable to adjudicate between people who disagree on both the rules of inference and their considered judgments about particular cases. One might also wonder whether this method can get us anywhere if there are no completely incontrovertible judgments about particular cases in science. It is not my task in this book to provide a rigorous

Physics, Structure, and Reality. Jill North, Oxford University Press (2021). © Jill North.
DOI: 10.1093/oso/9780192894106.003.0003

epistemology of science (let alone to defend the coherentist attitude to justification the method of reflective equilibrium seems to assume). I nonetheless hope to show that, given how familiar and widespread these kinds of inferences are, we should take the principles that underlie them seriously, and we should generally trust their results. Indeed, I suspect that this kind of inconclusiveness pervades the epistemology of science. No conclusive argument can be given to resolve the debate between someone who prizes simplicity above all else and believes that a particular scientific theory is simplest, and another who prizes explanatory power above all else and believes that a different theory is most explanatory. Even so, we manage to come to reasonable (if not incontestable) conclusions about various cases by relying on some generally accepted criteria and principles of theory choice.

The main thing for us is to see that many familiar inferences we make about structure in physics are guided by certain epistemic rules or principles, and that even though these principles do not yield conclusive results, they do generally yield conclusions we accept as reasonable, so that we reasonably take ourselves to be justified in adhering to those rules.

Above all, we put the notion of structure to use in figuring out the nature of the physical world. The rules concerning structure guide our inferences from the mathematical formulation of a physical theory to the nature of the world according to that theory. In the case of a physical theory, then, what we ultimately learn about by thinking about structure is the true nature of the physical world.

Of course, it is not as though we are handed a physical theory mathematically formulated from on high, left to theorize in a vacuum about what the world must be like, assuming it is the true theory. We devise the mathematical formulation of a physical theory in the first place on the basis of various pieces of evidence, empirical and theoretical. (There will also be some initial physical posits that play a role in choosing a mathematical formulation; I discuss this at points later on.) Still, once we formulate the theory in a particular way on the basis of that evidence, there is work left to be done in figuring out the full nature of the world. It is here that the structure principles step in to help guide our theorizing.

3.2 Structure presupposed by the laws

One general principle we rely on when theorizing about the nature of the world according to a physical theory is this: infer physical structure in the world from the mathematical structure presupposed by the laws. (Assuming in particular that these are fundamental laws; this will become clearer as we proceed.)

To see that we generally adhere to a principle like this, consider three examples. These are examples of inferring a particular spatial, temporal, or spatiotemporal structure from certain candidate fundamental physical laws.

First consider Aristotle's physics. (Aristotle's physics was not formulated mathematically, but the example will reveal, in a particularly clear manner, the reasons for positing a certain structure in the world on the basis of the fundamental physical laws, reasons that carry over to explicitly mathematically formulated theories. One can further imagine a mathematical formalization of this physics, which would lead to the same conclusions; compare Section 3.5 below.)

According to Aristotle, there are different kinds of material elements in the universe, each of which has a particular kind of motion that is natural for it, the kind of motion the element displays when unimpeded, so that it tends to move toward its natural place.

Importantly, for Aristotle, the universe as a whole is spatially spherical, with a distinguished center. This spatial structure is referred to in characterizing the natural motions of the elements. Heavy elements, like earth, naturally move downward, toward the privileged center, whereas lighter elements, like air, naturally move upward, away from the center.

By referring to the center of the spherical universe in characterizing the natural motions of the elements, Aristotle's physics presupposes that space has a spherical geometry, with a preferred central location. If space did not have a preferred-location structure, then this physics would not make sense: the basic principles of motion would refer to a privileged location that does not exist, that is not well-defined. (Although I have put this in substantivalist terms, none of this requires substantivalism, the view that space exists. The standard view is that Aristotle was a relationalist, and the relationalist can arguably also make sense of claims about the spatial structure of a world, or so I think, for reasons in Chapter 5. I will continue to speak in substantivalist terms here, for ease of exposition.)

Aristotle's physics cannot be stated without assuming that there is a preferred location, which the basic principles of motion invoke: the natural motions of the elements are defined by reference to it. In this sense, the laws of Aristotle's physics *presuppose* or *require* this structure. They cannot be meaningfully formulated without it.[1] (If you are starting to worry that this will depend on exactly how the laws are formulated, stay tuned until Section 3.5.)

And since these laws presuppose a particular spatial structure in their formulation, we infer that there is a certain physical structure in the world. We infer that, in a world fundamentally governed by Aristotle's physics, a world for which Aristotle's principles are the fundamental physical laws, space has a spherical geometry, with a privileged center. Indeed, one reason to think that Aristotle's physics is not the fundamental physics of our world is that we don't think our universe has the requisite spatial structure.

[1] There will be other structure required by this physics. I am focusing on one aspect for illustrative purposes.

Next consider Newtonian physics. Take Newton's first law, which says that an object continues with uniform velocity unless acted on by a net external force. In spacetime terms: an object continues on a straight spatiotemporal trajectory unless acted on by a net external force. (It is anachronistic to conceptualize Newton's physics in terms of spacetime rather than space and time, but it has become standard practice in philosophy of physics to do so, and I will generally follow suit. This does however raise some difficult interpretive questions, which I will touch on in Chapter 7.)

Newton's first law says that objects behave differently depending on whether they are traveling with uniform velocity or not; that is, depending on whether they are traveling on a straight spacetime trajectory or not. In this way, the law presupposes that there is a distinction between the straight spacetime trajectories, which represent uniform-velocity, inertial motions, and the curved trajectories, which represent non-uniform, non-inertial motions.[2] In other words, the law presupposes that spacetime has an inertial or affine structure, the kind of structure that distinguishes between straight and curved trajectories. Recall that an affine structure is needed to define a notion of straight as opposed to curved lines. Without such a structure, there simply would be no distinction between straight and curved spacetime trajectories: the notion of "straight" versus "curved" would be undefined, and a law like Newton's, which assumes the distinction—telling things to behave differently depending on which kind of trajectory they are on—would not make sense.

Newton's law cannot be formulated without assuming an affine or inertial structure; it presupposes it. Maudlin expresses the idea this way:

> In order for Newton's Law to *make sense*, for it to *make any claim at all*, there must be a distinction in nature between the trajectories of particles which are at rest or in uniform motion and those which are not. So there must be enough objective structure in space-time to found such a distinction.
>
> (Maudlin, 2011, 35; original italics)[3]

Therefore, in a fundamentally Newtonian world, we reasonably infer that there is an inertial spacetime structure. As in the case of Aristotle's physics, here, too, we see that the laws presuppose a certain structure in their formulation, which leads us to infer that the world has a particular physical structure. (Although it is standard to think that Newton's laws require an inertial structure—this is implicit in standard textbook presentations—this is controversial among philosophers of

[2] Strictly speaking the law assumes that there is such a distinction. We then add a natural assumption that the straight trajectories correspond to inertial motions and the curved ones to non-inertial motions, so that objects that are not accelerating relative to one another will all be on similarly straight trajectories.

[3] Compare Maudlin (2012, 10); Pooley (2013, 527).

physics, for reasons that have come to light in recent literature. I return to this in later chapters. For now, assume the usual viewpoint, based on a standard conception and formulation of Newton's laws.)

There is a slight difference between the cases of Aristotle's and Newton's laws. For there is a slight difference between a law's *mentioning* some structure and its *presupposing* it. Aristotle's laws explicitly mention a preferred spatial location, thereby presupposing this structure. Newton's first law does not explicitly mention the affine structure of spacetime, but it nonetheless presupposes it. This difference does not matter when it comes to making these inferences, however. Either way, we can see that the laws require some structure in order to be meaningfully stated, and we infer from this that the world has a certain physical structure.

Notice that Newton's law does not presuppose that space has a preferred location, by contrast to the principles of Aristotle's physics: there is neither explicit mention nor implicit presupposition of such a structure. In fact, none of Newton's laws require a privileged spatial location. One way to see this (briefly here; more in Sections 3.3 and 3.5) is to note that the mathematical form of the equations, the equations expressing the laws in mathematical language, does not change when we shift the spatial coordinate origin. The laws "look the same" or "say the same thing" in coordinate systems that are spatially translated relative to one another; they are invariant under changes in coordinate origin.[4] This reveals that the laws do not presuppose or require a preferred-location structure, for they make the same predictions, they tell things to behave the same way—they have the same mathematical form—regardless of which location we choose to be the origin. In the case of Newton's laws, unlike Aristotle's physics, we do not attribute a preferred-location structure to the world.

These two examples illustrate that we generally learn about physical structure in the world from the mathematical structure needed to support the fundamental laws—"support" in that the laws cannot be formulated, they wouldn't make sense, without it. As Earman puts it: "laws of motion cannot be written on thin air alone but require the support of various space-time structures" (1989, 46). The structure required to "write" the laws, in turn, then tells us about physical structure in the world. We infer that the world has a particular physical structure on the basis of the mathematical structure presupposed by the fundamental laws—as we infer that there is a certain spatial structure from the principles of Aristotle's physics and a certain spatiotemporal structure from Newton's laws. We likewise infer different physical structures from different requisite mathematical structures—as we infer that there is a distinguished spatial location from Aristotle's physics but not Newton's. (The examples further suggest that we don't ascribe to the world more structure than what's needed to support the laws: hold that thought until the following section.)

[4] Again, if you are concerned that this may depend on the particular formulation of the laws, be patient until Section 3.5.

Notice that when making these inferences, we assume that the laws in question are fundamental laws. Assuming that Aristotle's physics is the fundamental theory, we infer that the world has a preferred-location structure. Assuming that Newton's physics is the fundamental theory, we are led to ascribe a different structure to the world. (Perhaps there is some similar principle at work for nonfundamental laws, but I won't consider this here.)

Notice, too, that we make this type of inference on the basis of laws that are explicitly formulated in mathematical terms, as is the case with current formulations of candidate fundamental theories, as well as laws that are not so formulated, such as Aristotle's principles or the laws as Newton himself stated them. Either way, we can see that there is some mathematical structure the laws implicitly require, and on this basis we posit a physical structure in a world of which these are the fundamental laws. Note the two types of structure in play here: a mathematical structure required by the laws when thought of as mathematical equations, and a physical structure had by a world that those laws represent. (Aren't there often different mathematical formulations of the laws available? If the laws can be formulated using different mathematical structures, what physical structure do we infer to the world then? I turn to this question in the next chapter.)

You may wonder why we can learn about a world's physical structure in this way. The underlying idea is simple. Certain aspects of the world, such as the nature of space and time, can't be directly observed, nor can they be straightforwardly inferred from the phenomena. Things are not distributed uniformly throughout space, for example, but we don't infer from this alone that space has an asymmetric structure.

We can learn about these things in a different way: from features of the fundamental laws. If the fundamental laws cannot be formulated without assuming some structure, then plausibly that structure must exist in a world governed by those laws. This then gives us reason to infer that the world has the structure—that is, that the world has physical structure corresponding to the mathematical structure required to support the laws.

This idea is motivated by the realist commitment mentioned in Chapter 1, although there is a bit more to it. It is not just that the structure in question is required for the laws to be true; it is that it is needed for the laws to even make sense. It is hard to see how to make sense of a law which says that water naturally moves toward the center of the universe if there is no such thing as the preferred center of the universe. It is hard to see how to make sense of a law which says that objects depart from their straight spacetime trajectories in certain circumstances if there is no such thing as a straight as opposed to curved spacetime trajectory.

The rule to infer physical structure in the world from the mathematical structure presupposed by the fundamental laws may sound like Quine's (1948) criterion for ontological commitment, but it is not exactly the same. Quine says that we are ontologically committed to what the variables of our best theories must range over in order for those theories to be true. Quine's prescription has to do with ontology, with what entities exist. The current idea concerns what structure must be assumed in order to meaningfully state the laws. One difference from Quine is that the structure we attribute to the world needn't be explicitly quantified over in the laws, but can be implicitly assumed or presupposed. Another is that the requisite structure is not just needed for the laws to be true, but for them to be formulated in such a way as to be capable of having a truth value. Another is that the question of what entities exist can come apart from the question of what structure there is (we will see this in particular in Chapter 5).

Here is one final example of following the rule, in this case concerning the nature of time. Consider the notion of time reversal invariance, which is a symmetry of a law or theory that is unchanged when we swap past and future, inverting the direction of time. Time reversal invariant laws remain the same, they have the same form, under a change in time coordinate from t to $-t$. Alternatively (on an active understanding of time reversal), take a solution to a theory, a sequence or history of states that is possible according to the theory, and invert the time order of the sequence of states. A theory is time reversal invariant when this always transforms a solution into another solution.[5] Newton's laws are symmetric in this sense: any behavior allowed by these laws can also happen backward in time; passively, replacing t by $-t$ yields the same form of the laws.[6] Newton's laws do not distinguish between past and future; they don't recognize a difference between the two temporal directions. They say the same thing regardless of the direction of time.

By contrast, non-time reversal invariant laws do not say the same thing to the future as to the past. They say that different things can happen in one direction of time as opposed to the other. An example is the law of wavefunction collapse in (certain theories of) quantum mechanics, which assigns probabilities to the different possible wavefunctions that a system's current wavefunction could collapse into in the future. It does not assign probabilities to different possible wavefunctions in the past, given the current wavefunction. Such a law is not time reversal invariant.[7]

Non-time reversal invariant laws tell things to behave differently depending on the direction of time, thereby presupposing that there is a distinction between the

[5] This is eliding details and debate. See e.g. Albert (2000, Ch. 1); North (2008).

[6] Think in particular of Newton's second law, the central dynamical law of the theory: see Albert (2000, Ch. 1, esp. n. 6).

[7] As Arntzenius (1995) puts it, collapse theories are theories of forward transition chances but not backward transition chances. Price (1996, 2002a,b) argues that such a theory might in fact be symmetric in time, with backward transition chances that do not result in observed frequencies since they are "subordinate" chances that are overridden by the initial low entropy condition in that direction. However, quantum phenomena do not display invariant backward transition frequencies (Arntzenius, 1995, 1997). And it is not clear why we should believe in the existence of lawful chances in that temporal direction if they are never manifested in observable frequencies. More, as Arntzenius discusses, we cannot in general add invariant backward transition chances to a theory that has invariant forward transition chances, not without making the theory woefully empirically inadequate.

two temporal directions. Such laws assume or presuppose an asymmetric temporal structure, a temporal orientation, which is needed to define a distinction between the two temporal directions, in the same way that an affine structure is needed to define a distinction between straight and curved lines. (A temporal orientation is an everywhere continuous timelike vector field that contains only vectors that lie in one lobe of the light cones. It picks out, at each point, which direction is to the future and which is to the past of that point.[8])

In a world governed by fundamental non-time reversal invariant laws, therefore, we have reason to infer that there is a temporal orientation, just as we have reason to infer that a world fundamentally governed by Newton's laws possesses an inertial structure. Since non-time reversal invariant laws presuppose a temporal orientation, since this structure is needed to make sense of the distinction between the two temporal directions assumed by the laws, we reasonably infer that the world has a corresponding physical structure. Without a temporal orientation, after all, there wouldn't be a well-defined distinction between the two temporal directions, which makes it hard to see how the phenomena could exhibit a lawful difference in behavior between them.[9]

These three examples demonstrate that we generally impute or ascribe to the world the structure that's needed to support the fundamental laws. (They also suggest that we do not generally impute more structure than that: Section 3.3.) We infer that there is physical structure in the world corresponding to the mathematical structure presupposed by the fundamental laws, as when we infer that there is a privileged spatial location from the principles of Aristotle's physics; an inertial spacetime structure from the laws of Newtonian physics; or a temporal orientation structure from non-time reversal invariant laws. Overall, we take the mathematical structure presupposed by the fundamental laws *seriously* in that this tells us about physical structure in the world, and different such mathematical structures indicate different physical structures in the world.

These are not absolutely conclusive inferences based on an utterly infallible principle. There may be other considerations that point against positing a direction of time on the basis of non-time reversal invariant laws or an inertial structure on the basis of Newton's laws; and we will come across views that reject this type of inference altogether. These inferences are nonetheless reasonable and familiar. They are based on a plausible, ceteris paribus principle, which we are justified in adhering to in part because it yields inferences we take to be reasonable.

[8] A temporal orientation defines a distinction between the two temporal directions, though not which is future and which is past: there are two globally definable orientations.

[9] The idea that we should posit a temporal orientation on the basis of non-time reversal invariant laws is reasonably widespread. It can be found in Earman (1969, 2002); Horwich (1987); Arntzenius (1995, 1997); Callender (2000); Maudlin (2007b). (Maudlin however argues that not just an orientation is needed, but also the passage of time.)

3.3 Minimizing and matching structure

The above examples illustrate that we generally posit physical structure in the world on the basis of the mathematical structure needed to support the fundamental laws. In addition, we generally do not posit any more structure than that. In other words, we also adhere to a minimize-structure principle, which says to posit the *least* structure required for the fundamental laws—both mathematical structure in the formalism and physical structure in the world. (This rule helps indicate both what the right formalism is, and what that formalism says about the world. My focus in this chapter is on the latter aspect of the rule. I turn to the former aspect in Chapter 4.)

The minimize-structure rule holds ceteris paribus. All other things being equal, we should minimize structure, both in the formalism and in the world. There can be reasons to infer additional structure beyond what is needed to support the fundamental physical laws. In the absence of such reasons, we should not do so.

The minimize-structure rule was implicit in the previous section. Consider the thought that according to Newtonian mechanics, there is no privileged spatial location. Since Newton's laws do not recognize or pay attention to differences in coordinate origin, we infer that the spatial structure of a world fundamentally governed by these laws is homogeneous: the laws do not assume a preferred location, and we correspondingly infer that there is none. Or take the case of time reversal invariance. Newton's laws do not recognize a difference between the two temporal directions: they say the same thing regardless of the direction of time. This means that they don't presuppose the structure needed to distinguish between the two temporal directions. Therefore, in a world fundamentally governed by Newton's laws, we do not posit a temporal orientation: we infer that there is no "direction of time" in a fundamentally Newtonian world.

In this section, I want to focus on two different, particularly familiar and important, inferences that illustrate our reliance on the minimize-structure rule as well as its ceteris paribus character.

The first comes from Newtonian physics. Newton thought that his physics requires the existence of absolute space, a space that persists through time, relative to which there are absolute facts about objects' spatial locations and velocities (understood as changes in their absolute locations over time). He argued that phenomena involving inertial and non-inertial motion reveal this: consider the bucket experiment or the spinning globes example (more on these in Chapter 5). Putting this (anachronistically) in spacetime terms, Newton believed that his physics requires the structure of what has come to be called Newtonian spacetime. This spacetime possesses structure corresponding to Newton's idea of absolute space, for it has structure that identifies spatial locations over time.[10]

[10] This is also sometimes called Aristotelian spacetime (for instance by Geroch (1978)), though notice that it is different from the structure that Aristotle himself assumed, and which Earman (1989, Sec. 2.6) calls Aristotelian spacetime.

However, we now think that Newton was wrong about the structure required for this physics, and the reason comes from the minimize-structure principle.

To see this, first note that Newtonian spacetime has more structure (in various senses discussed in Chapter 2) than an otherwise-similar spacetime called Galilean (or neo-Newtonian) spacetime. Newtonian spacetime has all the structure of a Galilean spacetime, but it adds an absolute space structure, a relation of "occurring at the same spatial location" between events at different times, and a corresponding notion of absolute velocity.

Second, note that Newton's laws do not require this additional structure. They don't presuppose a distinction between being at rest and moving with constant nonzero velocity: they say the same thing, they make the same predictions, regardless of what (constant) velocity an object has. Newton's laws do not recognize differences in absolute velocity.

One way to see this, familiar from physics books, is to consider what happens to the laws under a change in, or transformation of, inertial reference frame. It turns out that Newton's laws have the same mathematical form when expressed in terms of the coordinates of any inertial reference frame. They have the same form when expressed in terms of the (x, y, z, t) coordinates of one inertial frame as they do in the (x', y', z', t') coordinates of any other frame related to the first by a shift in constant velocity, or uniform velocity boost.[11]

This is not hard to see. Take Newton's second law. The equation $\mathrm{F} = m\mathrm{a} = m\frac{d\mathrm{v}}{dt} = m\frac{d^2\mathrm{x}}{dt^2}$ (or $\mathrm{F} = m\ddot{\mathrm{x}}$, in the notation I use in the following chapter) will be unaffected by a uniform velocity boost, since each quantity appearing in the law is unchanged by the addition of any constant velocity. F and m are assumed to be independent of velocity, and will therefore be unaffected by the transformation.[12] Acceleration will also be unchanged. Suppose that **v** is an object's velocity in the unprimed frame, v' its velocity in the primed frame, and that the primed frame is moving with velocity **V** with respect to the unprimed frame, so that $\mathrm{v}' = \mathrm{v} - \mathrm{V}$, where **V** is any constant velocity. Then $\mathrm{a}' = \frac{d\mathrm{v}'}{dt} = \frac{d(\mathrm{v}-\mathrm{V})}{dt} = \frac{d\mathrm{v}}{dt} = \mathrm{a}$. (In classical physics, $t = t'$ for any two reference frames.)

Newton's equation will therefore have the same form in the primed frame that it did in the original. Given an inertial frame in which particles obey $\mathrm{F} = m\mathrm{a}$, in any other inertial frame particles will obey $\mathrm{F}' = m\mathrm{a}'$—the very same equation, the only difference being the primes on the variables. This demonstrates that the same law will be obeyed in the relatively moving frame as in the original one.[13] (What happens to this reasoning if we transform to non-rectangular coordinates? In polar

[11] This invariance in form is sometimes called the covariance of the laws, a term I avoid as much as possible since it is not used univocally and there are controversies surrounding it.

[12] The assumption about mass is made for any non-relativistic theory. The assumption about forces is also typical for any classical theory, although there are some reasons to question it, as will be noted in Chapter 4.

[13] This kind of argument can be found in just about any physics textbook that mentions the Galilean invariance (as it is called; below) of Newton's laws; two examples are Schutz (2009, Sec. 1.1) and Shankar (2014, Sec. 12.2). An analogous argument can be given on the basis of an active transformation: take

coordinates, for example, the second law does not have the form $\mathrm{F} = m\frac{d^2\mathrm{r}}{dt^2}$. I turn to this in Chapter 4.)

Newton's first law is also invariant under a uniform velocity boost, since if an object is moving with a constant velocity in one inertial frame, it will continue to do so after any constant velocity is added to its velocity, as effectively happens when we switch to another inertial frame. Assuming that forces are independent of velocity, as usually assumed on this theory, Newton's third law (as well as any particular force laws, such as the law of gravitation) will be unaffected by the transformation as well.

In all, Newton's laws are invariant under transformations in uniform velocity or inertial reference frame. They "say the same thing"—they make the same predictions, they have the same mathematical form—regardless of inertial frame. Since the transformation equations between inertial frames in Newtonian physics are called the Galilean transformation equations, we say that Newton's laws are invariant under Galilean transformations, or Galilean invariant. (These transformations include spatial translations and rotations as well as uniform velocity boosts.)

Although the idea of the mathematical form of an equation is, as one textbook puts it, "difficult to give a precise meaning to" (Hartle, 2003, 37), inferences such as the above are ubiquitous in physics (the quoted author himself going on to apply it to Newton's equations). For the purposes of this book, I am going to assume that this type of inference—an inference based on the invariance in form, under various transformations, of the equations expressing the laws—is justified, in particular given its ubiquity. I will say some more about this in Section 3.5.

Newton's laws say the same thing regardless of inertial frame; they do not recognize differences in constant velocity. This, in turn, means that we can formulate the laws without presupposing a preferred inertial frame and an underlying absolute space structure, the structure needed to support facts about absolute velocity. Help yourself to a privileged frame in formulating Newton's laws; transform to any other inertial frame, moving at any other constant velocity; and you will see that the laws are unaffected—they remain the same under the transformation. This reveals that the laws do not recognize or make use of such a structure. They do not presuppose a preferred frame and the absolute space structure that privileging such a frame requires.

And since this structure isn't needed to formulate the laws, we infer that there is no corresponding physical structure in the world. In a world fundamentally governed by Newton's laws, we infer that there is no absolute space structure, the kind of structure that would pick out one frame as being *really* at rest—at rest in an absolute, frame-independent sense. (As one physics book concludes, on the basis

any solution to the equations, apply a uniform velocity boost to it, and you always get another solution, another history that is possible according to the laws.

of the form-invariance of Newton's laws under the Galilean transformations: "So there is no absolute velocity" (Schutz, 2009, 2).)

In accordance with the minimize-structure rule, in other words, we infer that absolute space is excess structure beyond what is needed for the laws. We infer that a Galilean spacetime structure is the right structure to posit in a world fundamentally governed by Newton's laws, despite what Newton himself (or an anachronistic spacetime version of himself) thought. The idea that Newton's laws presuppose a Galilean spacetime structure is standard, if not always explicit, in physics as well as philosophy. (Newton was aware of the Galilean invariance of his laws, expressed in Corollary V of the *Principia*. He had other reasons, stemming from his metaphysics and theology, for thinking this physics requires absolute space. (Recall the ceteris paribus nature of the inference rules.) The mathematics available at the time further suggested the need for facts about absolute location and velocity in order for there to be facts about absolute acceleration. None of the empirical reasoning of the *Principia*, however, really requires that structure (Smith, 2008), and the usual view nowadays is that the Galilean invariance of the laws indicates otherwise.)

Newton himself stated the first law in a way that does seem to presuppose a distinction between absolute rest and motion: "Every body continues in its state of rest, or of uniform motion in a right line, unless it is compelled to change that state by forces impressed upon it" (1934, 13). This statement of the law explicitly mentions both "at rest" and "uniform motion" and therefore seems to require an absolute space structure in order to found the distinction. However, according to this formulation of the law, too, things behave in the same way regardless of whether they are at rest or in uniform motion—regardless, they persist in that state unless acted on by a net external force—which reveals that the law does not really presuppose a distinction between the two types of motion, despite superficial appearances to the contrary. It makes the same predictions regardless.[14]

(A deep question remains. If we adhere to Newton's own conception of velocity as a change of location in absolute space, and adopt Newton's space and time framework, then it seems as though absolute space is needed for the law to make sense after all. I am assuming a thought that is standard nowadays in philosophy of physics, which is that the above reasoning brings to light "not what Newton says about [the structure of space and time], but what is in actual fact presupposed by the science of dynamics that we associate with his name" (Stein, 1970a, 258). At the very least, this is one natural conception of Newton's physics. This does lead

[14] That said, there is a sense of "presuppose" on which it does presuppose facts about absolute velocity. I am focusing on the sense that is central to the above familiar inferences, encapsulated in the quotation from Stein in the next paragraph. This might lead you to seek a formulation that eschews any mention of velocity; one is in Maudlin (2012, Ch. 3).

to a question of why, and whether, this now-standard thought is reasonable. In Chapter 7, I turn to some subtleties behind the standard way of thinking.)

Given the minimize-structure rule, it is natural to wonder whether Newton's laws can get by with less than a Galilean spacetime structure. The above considerations suggest that they can't. Newton's laws distinguish between unaccelerated and accelerated motion (inertial and non-inertial spacetime trajectories), telling things to behave differently depending on whether they are accelerating or not (whether they are on an inertial trajectory or not). Newton's first law says that an object will continue with uniform motion, on a straight spacetime trajectory, unless acted on by a net external force; in which case the motion will become non-uniform, along a curved trajectory, in a way dictated by the second law. These laws pay attention to the difference between inertial and non-inertial motion. They make different predictions, they tell things to behave differently, depending on whether an object is on a straight or curved spacetime trajectory. They therefore require or presuppose the structure necessary to distinguish between the two kinds of motion or trajectory, namely, an affine or inertial structure, which Galilean spacetime possesses. This structure is required to state or "write" the laws, as Earman puts it.

We want to eliminate structure whenever possible, in other words, but not beyond what's truly required by the fundamental laws. Galilean spacetime seems to be "exactly the structure that Newtonian dynamics requires" (Maudlin, 1993, 192), since it supports the requisite quantity of acceleration, without any excess structure. (Some philosophers have argued recently that Newtonian physics can get by with less structure (Saunders, 2013; Knox, 2014). Since my aim is to illustrate our reliance on the minimize-structure rule, and those arguments also implicitly rely on such a rule, for now I assume the more standard inference. We will see later on that those who argue for a different structure have a different conception of Newton's laws, which results in their differing conclusions about the requisite structure on the basis of similar principles.)

Notice how the above reasoning assumes that we can compare Galilean and Newtonian spacetime with respect to their relative amounts of structure, in case you were harboring any skepticism about the kinds of structural comparisons discussed in Chapter 2. The two spacetimes have a lot of structure in common (they are both four-dimensional differentiable manifolds with the same topology, the same differentiable and affine structure, and with an absolute temporal metric and a Euclidean spatial metric on each simultaneity slice). But there is a clear sense in which one of them has more structure than the other. Newtonian spacetime possesses structure that picks out one of the class of straight trajectories specified by the affine structure as preferred, representing particles that are at rest with respect to absolute space. Galilean spacetime does not have this further structure. (Friedman's (1983, Secs. 3.1–3.2) discussion makes this particularly clear. Newtonian spacetime has all the geometric objects defined on it that Galilean spacetime does, but it also has an additional vector field, corresponding to a preferred "rigging" or privileged class of timelike geodesics, which picks out a preferred inertial frame. It is then easy to see that Galilean spacetime is "simply Newtonian space-time without the rigging" (Friedman, 1983, 87).) Newtonian spacetime has more structure in the variety of senses discussed in Chapter 2. More notions are defined or meaningful (absolute rest and velocity, absolute location); more facts or distinctions are recognized (being truly at rest versus being in motion, remaining at the same place versus changing location); additional mathematical objects are defined (the rigging that defines the preferred inertial frame). Thus, behind the familiar inference of a Galilean spacetime for Newton's laws lies the principle to minimize structure, and behind that principle lies the tacit assumption that we can compare different spacetimes with respect to their relative amounts of structure.

Turn now to another example illustrating our adherence to the minimize-structure rule. In special relativity, we infer that spacetime lacks an absolute simultaneity structure. Change from one inertial reference frame to another—from a frame in which one set of spacelike hypersurfaces are simultaneity planes (each such plane comprising the events taken to be simultaneous with one another) to another frame in which a different set of spacelike hypersurfaces are simultaneity planes—and the laws always remain the same. (Recall from Chapter 2 that the inertial frames are now the Lorentz frames, with coordinates that transform according to the Lorentz (or more generally, the Poincaré) transformations, not the inertial frames of Newtonian mechanics, which transform according to the Galilean transformations. One way to see this is from the fact that the equations expressing the laws retain their form under the Lorentz but not the Galilean transformations: they are Lorentz invariant, not Galilean invariant.) The dynamical laws, Maxwell's equations and the Lorentz force law, have the same form when expressed in terms of the coordinates of any one Lorentz frame as they do in the coordinates of any other, even though different Lorentz frames disagree on which sets of events are simultaneous.

These laws are invariant under changes in inertial frame. They make the same predictions, they "say the same thing," regardless of choice of Lorentz frame, and so regardless of which events are taken to be simultaneous. These laws do not recognize differences in absolute simultaneity; they don't recognize facts about which events are *really* simultaneous with one another. This, in turn, means that we can formulate the laws without implicitly referring to a preferred simultaneity frame and the absolute simultaneity structure that would underlie a preferred frame. Since a spacetime with an absolute simultaneity structure has more structure than an otherwise-similar spacetime without (in various ways discussed in Chapter 2), and since the laws do not require this structure in their formulation, we infer, in accordance with the minimize-structure rule, that a world fundamentally governed by these laws does not possess any such physical structure. Absolute simultaneity is excess structure beyond what's needed for the laws.

This reasoning is familiar. It is what underlies our preference for Einstein's theory of special relativity over Lorentz's ether theory, for instance. These theories have the same dynamics. Yet whereas Einstein's theory does away with absolute simultaneity, Lorentz's theory posits absolute, frame-independent facts as to which events are simultaneous, namely those that are simultaneous according to the rest frame of the ether. Lorentz's theory posits more spacetime structure than what is required by the dynamical laws. So we infer that, in a world fundamentally governed by these laws, spacetime has the structure assumed by Einstein's theory; and more generally we infer that Einstein's theory is preferable to Lorentz's theory (ceteris paribus). This is the conclusion that has generally been drawn in physics and in philosophy.[15]

This inference is not conclusive. There is room for debate over the existence of an absolute simultaneity structure and a preferred simultaneity frame,[16] as there is room for debate over an absolute space structure for Newtonian mechanics. The inference is nonetheless reasonable and familiar. All other things being equal, given the special relativistic laws, assuming these are the fundamental physical laws, we infer that spacetime does not have an absolute simultaneity structure. Things might not have been equal: we might have found empirical evidence of Lorentz's ether, for instance. As it happens, this was not the case, and we reasonably prefer Einstein's theory.

Notice how this inference, too, presupposes that we can compare different spacetimes with respect to their relative amounts of structure. There is a clear sense in which the spacetime of Lorentz's theory has more structure than that of Einstein's theory: it possesses an absolute simultaneity structure, which the spacetime of Einstein's theory lacks. Thus behind this inference, too, lies the principle to minimize structure as well as the tacit assumption that we can compare spacetimes in terms of their relative amounts of structure.

I should mention that in this case, the comparison of structure is not quite as straightforward as it was for the two versions of Newtonian mechanics. This is because Lorentz's ether theory does not contain extra structure in the very same way that a Newtonian spacetime version of Newton's physics contains extra structure compared to a Galilean spacetime version. It is natural to regard Lorentz's theory as espousing a Newtonian spacetime structure (even though the laws are Lorentz invariant, and even though Lorentz himself did not put it this way: this is an anachronism in the same way that a spacetime version of Newton's physics is). And removing the absolute space structure of Newtonian spacetime does not yield the Minkowski spacetime structure of Einstein's theory, but a Galilean one.

Nevertheless, there is still a clear sense in which Lorentz's theory assumes or requires more structure than Einstein's theory: it assumes an absolute simultaneity

[15] This is not to deny that there are other, closely related reasons that many have preferred Einstein's theory, such as its superior explanatory power/unification: Janssen (2002a,b, 2009).

[16] Especially when quantum mechanics enters the picture, as certain theories of quantum mechanics seem to require a preferred simultaneity frame.

structure, or absolute simultaneity facts, that Einstein's theory lacks, while having the same dynamics and saving the same phenomena. The spacetime structures of the two theories differ in a variety of ways from Chapter 2. So we can, and generally do, still apply the minimize-structure rule in preferring Einstein's theory to Lorentz's. This is one example I had in mind when I said, at the end of Chapter 2, that these comparisons of structure are useful even though they may not be completely clear-cut in every case. We should not throw away the usual judgment in this case just because it is hard to give an absolutely precise account of comparative structure. Physicists and philosophers of physics generally (if not universally[17]) prefer the non-Lorentzian conception of special relativity, and they are plausibly reacting to its lesser structure. That said, there are arguably ways of making the comparison precise in this case as well. For example, Thomas Barrett (2015b) shows that one means of comparing spacetime structures, by comparing the sizes of the automorphism groups—a measure of structure in keeping with the discussion of Chapter 2—yields a precise sense in which Newtonian spacetime possesses additional structure compared to Minkowski spacetime.[18]

(All of this raises a question: whether Lorentz's theory counts as relativistic, deserving of being considered a theory of special relativity—whether Einstein's and Lorentz's theories are two different versions of "special relativity." The question arises because even though on both theories any inertial observer will measure the same constant speed of light, in accordance with a core idea of special relativity, according to Lorentz's theory light in fact only travels at that speed in the absolute rest frame of the ether, which runs counter to the spirit of the principles of special relativity. In David Albert's phrases,[19] Lorentz's theory is not metaphysically Lorentz invariant, even though it is observationally Lorentz invariant. And you might think that only a metaphysically Lorentz invariant theory, like Einstein's, qualifies as a genuine theory of special relativity. We do not have to settle this here. (It depends on what we take to be the key principles of a given theory, and there may not always be a fact of the matter about this.) Regardless, Lorentz's theory contains additional structure compared to Einstein's, while having the same dynamical laws. This is all that is needed for the above inference.)

I have been presenting Lorentz's and Einstein's theories as differing in their spacetime structure, in line with a standard take on the theories. By contrast, on the dynamical approach to spacetime of Harvey Brown (2005), and in a different way the approach of Albert (2019b, 2020), a world's spacetime structure is nothing more than a statement about the dynamics: there is nothing to spacetime structure

[17] Brown (2005) is one notable exception.

[18] See Wilhelm (2021) for a suggested revision to Barrett's criteria. Bradley (2019) also says that Lorentz's theory has "strictly more structure" than Einstein's, and not in the sense that we can simply remove the preferred state of rest of Lorentz's theory to obtain a theory that is structurally equivalent to Einstein's, for we must also fiddle with the notion of simultaneity.

[19] From a seminar discussion at Rutgers in Spring 2017.

beyond the dynamics. According to this kind of view, there isn't any real difference between Einstein's and Lorentz's theories, since they agree on the dynamics. I do not have a conclusive argument against the Albert–Brown type of view. It simply denies one of my main starting points, the basic thought that the dynamical laws "require the support of various space-time structures," in Earman's phrase, so that the laws come apart from those structures. It seems to me that this starting point is more in line with our usual (realist) inferences in physics. It is what underlies the inferences discussed in Section 3.2 and in this one. It is what underlies the thought that Newtonian mechanics *can* be set in a Newtonian spacetime even though it is *better* set in a Galilean one, as well as the usual reasons for preferring Einstein's theory to Lorentz's.

Steven French (2014, Sec. 2.5) objects to the minimize-structure rule on the grounds that when we examine actual cases from the history of science, the extra structure often turned out to be fruitful, so that we should never have done away with it in the first place. However, the examples he gives are cases in which the extra structure either turned out not to be truly excess, or was merely heuristically useful. The first kind of case simply shows that we can be wrong in our inferences about what is excess rather than genuine structure; yet surely the reasonableness of following the rule does not require infallibility. The second kind of case shows that extra structure can indeed be useful; but this pragmatic benefit does not suffice to show that it corresponds to genuine structure in the world. (Compare the potentials in classical electromagnetism, which are not taken to correspond to real physical things in the world, despite their usefulness; more in Chapter 7.)

Other things being equal, we ascribe to the world the structure presupposed by the fundamental laws—we attribute physical structure to the world corresponding to the mathematical structure needed to support the fundamental laws—and we do not impute any more structure than that. Putting these two ideas together, we generally adhere to a *matching principle*, which says to ascribe physical structure to the world that matches or corresponds to the mathematical structure required by the fundamental laws. All things being equal, there should be a match in structure between the laws and the world. Theories obeying such a principle are "well-tuned" to the world, to borrow a phrase that Earman (1989, Ch. 3) uses for a somewhat different idea, discussed in Section 3.4. I will make particular use of the matching principle in Chapter 5.

A few final points on these inferences and the epistemic rules governing them. First, keep in mind that the laws involved in these inferences are candidate fundamental laws. When inferring things about the world's physical structure in these ways, we assume that the laws in question are fundamental.

Second, the laws involved in these inferences are typically dynamical laws, laws of motion, which govern systems' behavior over time. (Some philosophers think this is the only kind of law that there is.) For this reason, when it comes to discussing "the structure presupposed by the laws," I will sometimes refer to this as a theory's dynamical structure, where by this I mean alternately the mathematical structure needed to formulate the fundamental dynamical laws, or the physical structure this indicates in the world.[20] (I have not given an argument that dynamical laws are the ones to guide these inferences, but will simply note that this is the kind of law that centers in the inferences I have been discussing. I leave it open whether similar things could apply to non-dynamical laws or principles.)

Third, notice that nothing in this discussion requires the so-called semantic view of theories, according to which a scientific theory is identified with its set or class of models. (More on this point in Chapter 6.) This is as opposed to the syntactic view, on which a scientific theory is identified with a set of sentences or axioms. It may seem as though my emphasis on a theory's structure, especially my emphasis on taking a theory's mathematical structure seriously in indicating physical structure in the world, rests on a semantic conception of theories. Many structural realists (in the sense from Chapter 1) do gravitate toward the view. Ladyman, for one, says that the semantic view, "which is to be preferred on independent grounds, is particularly appropriate for the structural realist. This is because the semantic approach itself contains an emphasis on *structures*" (1998, 416; original italics). However, the discussion in this book is orthogonal to that debate, allowing for any conception of scientific theories you might prefer. I do focus on the mathematical structures of our physical theories and what this tells us about the physical world, but in such a way as to remain neutral on what a scientific theory is. Even proponents of the syntactic view, after all, can talk about the mathematical structure required to formulate a theory, as well as the physical structure this may lead us to posit in the world.

Fourth, as I have said before, I take the discussion in this book to be predicated on a thoroughgoing scientific realism. It seems to me that the above principles are especially natural for the realist to accept. That said, these principles can appeal to a certain stripe of antirealist. Someone like Ruetsche (2011), recall from Chapter 1, can grant the reasonableness of these inferences, can allow that the above principles help us figure out what a candidate fundamental theory is saying about the world, while at the same time refraining from believing what the theory is saying. However, I am going to continue on the assumption of scientific realism.

Finally, let me reiterate that these principles yield non-conclusive inferences. This shouldn't trouble us. Compare any of our usual criteria of scientific theory choice—simplicity, explanatory power, and the like. These criteria are often imprecise; it can be unclear which to weight most heavily when their verdicts collide; and these verdicts only hold ceteris paribus. We do not choose the simplest theory no matter what, for example, but the simplest theory, ceteris paribus. (We do not

[20] This is not to be confused with a different idea. Stachel distinguishes the spacetime structures of a theory from its dynamical structures, where the latter "characteriz[e] the behavior of physical fields and/or particles in spacetime" (1993, 135).

stop with pure logic, despite its simplicity, as this would lack sufficient predictive power.[21]) We nonetheless take these criteria, and the principles governing them, to be reasonable and generally successful guides to theory choice; we routinely rely on them and do not feel epistemically at sea. Compare Glymour (1977) on the possibility of the underdetermination of a theory of spacetime geometry by empirical evidence. He argues that there won't in general be such cases of underdetermination, on the grounds that we will typically be able to compare the theories, and to choose between them, on the basis of extra-empirical criteria. He goes on to say of this idea:

> It is true that the principles of comparison are vague, and further that there is no principle given that determines which of these considerations take precedence should they conflict, or how they are to be weighted. (I doubt that there are any principles of this kind which are both natural and explain pervasive features of scientific practice.) I think there is rigor enough, however, to distinguish unambiguously among candidates that are offered in demonstration of the underdetermination of geometry. (Glymour, 1977, 236)

There is likewise rigor enough in our inferences concerning structure and the epistemic principles governing them. These are reasonable guiding principles, which we take to be generally successful, even though there may be cases for which the notion of comparative structure they rely on is not absolutely clear-cut, and even though their verdicts only hold ceteris paribus in particular. There could be reasons to posit an absolute space structure in a world fundamentally governed by Newton's laws (as Newton himself thought) or an absolute simultaneity structure in a special relativistic world (as Lorentz thought); in the absence of such reasons, we should not do so. To put it another way, anyone who advocates an absolute simultaneity structure in special relativity or an absolute space structure in Newtonian mechanics is saying that other things are not equal, and must argue as much. None of this means that our usual inferences are deficient.

3.4 Other principles

The above principles are implicit in many inferences we familiarly make in physics and philosophy of physics. Let me say a bit about how these principles are similar to, but not quite the same thing as, a few others that have been explicitly mentioned in the philosophical literature.

[21] Thus Lewis (1973, 73): "The virtues of simplicity and strength tend to conflict. Simplicity without strength can be had from pure logic, strength without simplicity from (the deductive closure of) an almanac."

One well-known principle in the vicinity is discussed by Earman (1989, Ch. 3). Earman says that the dynamical symmetries of a theory should be the same as the theory's spacetime symmetries. He spells this out in terms of two sub-principles—that any dynamical symmetry of a theory should be a spacetime symmetry of the theory, and that any spacetime symmetry should be a dynamical symmetry—and gives different arguments for each one.

Earman's symmetry principles will yield the same verdicts on the above cases. For example, the Galilean invariance of Newton's laws will indicate a Galilean spacetime structure, since in that case the dynamical symmetries will align with those of the spacetime structure. Not so if we were to posit a Newtonian spacetime structure: the spacetime would possess a different group of symmetries from the dynamics.

There are nonetheless two important differences in conception and emphasis between Earman's principles and the ones I have been discussing. The first is that the structure principles I have in mind are not "conditions of adequacy on theories of motion," as Earman (1989, 46) says of his principles, but guiding epistemic principles that hold ceteris paribus. Newton himself was not proposing an inadequate theory, as Earman's principles would have us believe. Rather, other things being equal—as they seem to be in this case, despite Newton's own reasons for thinking otherwise—we should not posit the structure he suggested. Similarly, Lorentz's ether theory is not beyond the pale. It is simply less preferable, other things being equal, despite Lorentz's reasons for thinking otherwise.

Similar considerations tell against Wayne Myrvold's (2019) idea that it makes no sense to talk about a spacetime structure that plays no role in the dynamical laws, an idea according to which Earman's principles are not just true, but analytically true, true by virtue of considerations of meaning. (Myrvold suggests this is how to understand the dynamical spacetime approach of Brown (2005).) On this view, it is not possible for a theory's spacetime symmetries and dynamical symmetries to come apart, for dynamical symmetries are all that we mean by spacetime symmetries, and vice versa. To say that a spacetime is Minkowskian and that the dynamical laws are Lorentz invariant, for example, is "to say the same thing" (Myrvold, 2019, 140).

I disagree. Lorentz's theory, with dynamical laws that are Lorentz invariant even though the spacetime structure is not Minkowskian, is not making a mistake about *meaning*. The theory makes sense; it is conceptually coherent. We understand what the theory is saying when it claims that lengths shrink and time dilates in just the right ways to prevent us from ever detecting the preferred rest frame of the ether, for instance. Nor was Newton talking nonsense when he posited the existence of an absolute state of rest, and an absolute space structure underlying it, that in fact played no dynamical role in his theory.[22] The reason we should not

[22] There is room for Myrvold to agree with this, if his view is limited to the structure of spacetime rather than space and time. However, he does not seem to want to limit his view in this way, since he uses the example of Newton's own physics to motivate the claim that these principles are analytic.

adopt these theories isn't that they "fail to make sense; they do not succeed in expressing anything at all, either true or false" (Myrvold, 2019, 137). The reason is that they exhibit an "epistemic vice," in the phrase of Shamik Dasgupta (2016).[23] Other things being equal, we should not infer structure that plays no role in, that is beyond what is needed for, the fundamental dynamical laws.

Myrvold is simply assuming that there is no non-dynamical conception of or underpinning to spacetime structure, and that any view saying otherwise is false. This is first of all contentious, as evidenced by the existence of alternative views, including realist but non-dynamical conceptions of spacetime structure like my own, but also views that take spacetime structure less seriously than I do. Reichenbach (1958), for instance, says that a world's spacetime structure is a matter of conventional choice. Yet in his view different such choices are possible on the basis of the dynamical laws: no particular choice is forced on us by the dynamics, let alone by virtue of considerations of meaning. Second, even if Myrvold's assumption is true, surely it is true because of what the physical world is like, and not just because of our meanings or concepts. We may well "ascribe structure to spacetime... on the basis of dynamical considerations, and shifts in dynamics and shifts in spacetime structure go hand in hand" (Myrvold, 2019, 139), but it does not follow that these shifts happen as a result of "conceptual analysis" (2019, 141). Rather, they happen as a result of shifts in our evidence for what the world's spacetime structure is.

There is a second difference between Earman's principles and the principles I defended above, which is Earman's explicit focus on symmetries. This is to some extent just a difference in emphasis, but it is important.

Recall that, intuitively and informally, symmetries are operations you can do to something so that it "looks the same" afterward, as a circle is symmetric under rotations. This intuitive idea, familiar for physical objects and geometric shapes, can be extended in a natural way to the physical laws. A symmetry of a law is something we can do to the law, or to an expression representing the law, so that it remains the same or "looks the same" afterward.[24]

We have seen that one way to figure out what structure is needed for the laws is to examine certain symmetries, namely the (passive) transformations that preserve the laws' form, or the (active) transformations that map solutions of the laws to solutions. Symmetries, especially dynamical symmetries, play an important role in our attempts to figure out what structure is required by the laws—enough so that a symmetry principle which says to align the spacetime symmetries with the dynamical symmetries, and a structure principle which says to posit physical

[23] Dasgupta is talking about a different but related principle having to do with undetectable structure, discussed later in this section.

[24] See Feynman (1965, Ch. 4); Brading and Castellani (2007) for more on this idea.

structure in the world corresponding to the mathematical structure needed for the dynamical laws, may seem to hardly differ.

There is a reason for formulating things in terms of structure rather than symmetries, though. Structure is what we are ultimately after (both mathematical structure in the formalism and physical structure in the world), and symmetries are simply an (important) guide to that structure. As mentioned in Chapter 2, symmetries are an indicator of structure, not the structure itself. More importantly, there can be more to the requisite structure than what seems to be indicated by dynamical symmetries.

As an example, in physics it is often said that different solutions to an equation that are related by a symmetry are physically equivalent. But as Gordon Belot points out, this idea, if unconstrained, "is a disaster," for "it will be possible to relate any pair of solutions by a symmetry" (2003, 402). If symmetries are simply transformations that map solutions of a theory onto solutions, then an arbitrary permutation of solutions will count as a symmetry of a theory. Yet this symmetry should not lead us to conclude that all solutions, to the equations of any theory, are physically equivalent—that any one is of them is an equally good representation of what is going on physically.[25] Or consider another example from Belot (2013, 330). Take a linear homogeneous partial differential equation, like the wave equation, for which adding a solution to any other solution always yields another solution. Belot notes that any two solutions to such an equation will be related by a symmetry. Combined with a familiar principle that things related by symmetries are identical or at least physically equivalent, a principle that Hilary Greaves and David Wallace say has achieved "widespread consensus" (2014, 60), such a symmetry would lead us to conclude, of any law mathematically represented by a linear equation, that all solutions are physically equivalent, if not identical. But surely we should not conclude that. After all, the law presupposes that there *are* physically distinct solutions, telling things to behave differently depending on the system and initial state. Taking symmetries as our sole guide to the requisite structure in these cases would be woefully misleading.

Since Earman's principle concerns the connection between dynamical symmetries and spacetime symmetries, it may be silent about inferences that don't explicitly involve spacetime symmetries, such as the above inferences concerning the equivalence of solutions. However, given the above-mentioned consensus, it is a dangerously small step from dynamical symmetries to the physical equivalence of different solutions, and from there to untoward repercussions for a theory's spacetime structure. More generally, the mention of symmetries in such a guiding

[25] Ismael and van Fraassen note, in a similar vein, that, "*It would in general make no sense to suggest that only features preserved by (invariant under) the symmetries of a given theory are real.* For to identify all possible worlds related by such symmetries (as representations of the same physical possibility) would be to claim that there is only one physically possible world" (2003, 378; original italics).

principle can result in a kind of tunnel vision, leading us to take dynamical symmetries as the sole basis for our physical reasoning, as Belot's (2013) discussion makes clear. This might lead you to go so far in the other direction as to reject any symmetry-to-equivalence reasoning.[26] Belot argues that this type of reasoning is at best overly simplistic, and I agree; but the important upshot for us is the reminder to refrain from focusing on symmetries per se in our physical theorizing. As we can see from the above examples, paying attention to what the laws presuppose is the better guide to the requisite structure, which symmetry-based reasoning fails to reveal. (To put the point in a way that foreshadows the discussion in Chapter 7: a mathematically definable symmetry need not thereby correspond to a physical symmetry or equivalence, another hint that we must be careful not to focus too narrowly on symmetries in our physical theorizing.)

In all, symmetries are an important tool to use in our theorizing about structure. As we will see further in Chapter 4, there is a particularly important type of symmetry to pay attention to in our inferences concerning structure, namely the invariance in form of the dynamical equations under various transformations. But this tool is used in service of a larger goal: to figure out what structure is needed for or presupposed by the fundamental physical laws. That is the ultimate goal, and our epistemic principles are properly formulated in terms of it.

Eleanor Knox endorses a different guiding principle: "one should, whenever possible, avoid postulating physical structure that is underdetermined by the empirical content of the theory" (2014, 866). This principle is not formulated in terms of symmetries. (She says that we should avoid Earman's formulation in terms of symmetries since what are a theory's dynamical versus spacetime symmetries can be at issue, as evidenced by the case of Newtonian gravitation she discusses.) Knox's principle, like Earman's, will result in similar conclusions for the above cases. Yet it, too, has a different focus, and suggests a different reason behind our inferences concerning structure.

Knox's principle codifies a widespread idea. Consider the inference to a Galilean rather than Newtonian spacetime structure for Newton's laws. Philosophers commonly say that the reason for this inference is that a Newtonian spacetime structure would give rise to in-principle undetectable physical facts. Newtonian spacetime possesses an absolute space structure, which picks out a preferred rest frame. Yet Newton's laws entail that no experiment could ever detect which is the preferred frame, since they make the same predictions regardless of inertial frame.[27]

[26] Alternatively, you might try to beef up the notion of symmetry while preserving the connection to equivalence. Ismael and van Fraassen (2003) argue that the genuine symmetries must not only map models to models but must also preserve any qualitative, perceptible features. Belot (2013) considers a variety of potential precisifications of the notion of symmetry that might avoid the concern, but ends up pessimistic that any of them will work.

[27] This idea is often mentioned in physics books (Feynman et al. (2010, Sec. 15.1) and Shankar (2014, Sec. 12.1) are two examples) even though, as Roberts (2008) and Dasgupta (2016) point out, a detailed argument for it is rarely given. Their discussions fill in missing details.

Therefore, the thought goes, we should infer a Galilean spacetime structure, which does not yield such in-principle undetectable facts. This is the verdict that Knox's principle yields, and the reasoning behind it is endorsed by many philosophers of physics.[28]

Knox's principle is fine as far as it goes. I agree that, other things being equal, we should avoid in-principle undetectable physical facts. However, I think that something more fundamental is going on in these inferences. The reason for the inference to a Galilean structure for Newton's laws is not just that no experiment could detect the additional structure of Newtonian spacetime (even though that is true). The reason is that positing no more structure than what is needed for the laws is *evidence* that we have inferred the *correct* structure to a world governed by those laws. This is a more fundamental reason for the inference than the verificationist-style desire to avoid undetectable physical facts. It is this reason that motivates the above principles.

Finally, it is worth mentioning that the structure principles do not amount exactly to Occam's razor, which concerns theoretical simplicity or ontological parsimony, although they are in a similar spirit. The minimize-structure rule in particular does not just say to posit the fewest entities or to infer the simplest theory, but to posit the least structure needed to support the fundamental dynamical laws. Occam's principle could be spelled out in terms of structure, I suppose, in which case the structure principles may seem to express a similar sentiment. But they are not exactly the same thing, or in any case they needn't be. (This is not a deep dispute. If you want to think of the structure principles as variants of Occam's dictum, I won't vehemently object.)

3.5 Invariance, structure, and coordinates

Consider the inference of a Galilean spacetime structure for Newtonian physics. One way to see that we should infer this structure is to note that Newton's laws make use of quantities that are well-defined in such a structure, like acceleration, and they do not make use of quantities that are not well-defined in it, like absolute velocity, as evident from the invariance of the laws under changes in inertial frame. Newton's laws say the same thing regardless of which inertial frame we choose, and so regardless of what (constant) velocities objects have. They therefore do not presuppose the structure necessary to support facts about absolute velocity.

[28] Some examples: Earman (1989, Ch. 3), Ismael and van Fraassen (2003); Roberts (2008); Dasgupta (2009, 2016); Maudlin (2012, Ch. 3); Pooley (2013, Secs. 3–4).

We saw this above, but there are some additional details and complications I wish to turn to now. This will lead to further notes on the use of coordinates in physics and the idea of an equation's form, some of which we were starting to see in Chapter 2.

The first thing to note is that our inferences from the laws to underlying structure—in particular, our inferences based on whether the laws "say the same thing," in the sense of being invariant or having the same form under various transformations—are sensitive both to how we state the laws and how we understand the transformations used to test for the laws' invariances. This might make you worry about the reliability of these inferences. It is important to realize, however, that none of this interferes with the basic legitimacy of the reasoning. It won't affect our conclusions about any case.

To see this, consider a simple example. Take a physics that is like Aristotle's in mentioning a certain location toward which different kinds of elements naturally move, but differs in that space is infinite in extent rather than finite and spherical. (Again assume substantivalism for ease of discussion.) Suppose this physics includes the principle that "the element earth naturally moves toward spatial location p." This principle explicitly refers to a particular spatial location. It thereby presupposes a preferred-location spatial structure. Here it is easy to see that this structure is required by the physics.

Let's see if considering whether the principle says the same thing under various transformations also yields this conclusion. Suppose we choose a coordinate system, S, in which p is located at the origin. When expressed in terms of this coordinate system, the above principle says that "the element earth naturally moves toward the origin $(x_0, y_0, z_0) = (0, 0, 0)$."

Now imagine another coordinate system, S', that is spatially translated three units to the right relative to S. The origin of S' is $(x'_0, y'_0, z'_0) = (0, 0, 0)$. In terms of the coordinates of S, the origin of S' is $(x_0, y_0, z_0) = (3, 0, 0)$.

In terms of coordinate system S', the principle stated above will no longer be true: it is not true that "the element earth naturally moves toward the origin $(x'_0, y'_0, z'_0) = (0, 0, 0)$," for this point does not correspond to the privileged location p. Instead, a spatially translated version of the principle will be true: "the element earth naturally moves toward a location three units to the left of the origin, $(x', y', z') = (-3, 0, 0)$." That said, it does remain true, in coordinate system S', that earth naturally moves toward point p. It is just that p is not at the origin of this coordinate system.[29]

[29] I am assuming that it makes sense to keep the privileged point p fixed under the transformation, so that the element earth no longer moves toward p after the transformation but toward another location, q. Someone could object that the way to *identify* point p in the shifted case is by means of the location with respect to which the element earth exhibits its natural motion. The above brings out the complications I wish to address here, but there are clearly further metaphysical debates to be had.

So we seem to have a bit of a mess on our hands. In one way, the original principle does "remain the same" or "say the same thing" regardless of the shift in coordinate system: earth naturally moves toward point p regardless of which coordinate system we use. But in another way, the principle does not "remain the same" or "say the same thing" regardless of the shift in coordinate system. Given a statement of the principle in terms of coordinates, the original principle does not remain true after the spatial translation, and in that sense it does not "stay the same" or "say the same thing"; instead a different, shifted principle holds. Another way to put it is that the principle does not retain its form. Before the coordinate transformation, the following principle holds: "earth naturally moves toward (x_0, y_0, z_0)." After the coordinate transformation, the following principle does not hold: "earth naturally moves toward (x'_0, y'_0, z'_0)." The very same form of principle, differing only in the primes appearing on the variables, no longer holds. (Compare the inference in Section 3.3 based on the Galilean invariance of Newton's laws. There we paid attention to the fact that the form of the laws remained the same under the relevant transformation, the only difference between the original and transformed equations being the primes appearing on the variables.)

Expressed in a fully coordinate-independent way, without any mention of coordinates, as the statement that "the element earth naturally moves toward point p," the principle is unaffected by a shift in coordinates. Expressed in terms of coordinates (and in a way that is coordinate-dependent, not just coordinate-based, in terms of the distinction mentioned earlier), as "the element earth naturally moves toward the origin," the principle is affected by the coordinate transformation. Whether the principle stays the same, says the same thing, or retains its form under a transformation depends on how we formulate the principle to begin with. Once again, this makes it hard to see how we can draw substantive physical conclusions by examining what happens to the form of the laws under various transformations: surely our conclusions about the nature of the physical world should not depend on how we choose to formulate the laws. On the other hand, drawing physical conclusions on the basis of the form of the laws is exactly what we seemed able to do for the equations of Newtonian mechanics and special relativity.

Notice, though, that if we were to apply an active transformation rather than a passive coordinate transformation, then the coordinate-independent formulation of the principle will be altered. Spatially shift objects' locations throughout history three units to the left. (This is the shift that yields the same changes to objects' coordinate-dependent descriptions that the above shift in the coordinate system did.) It will no longer be true, after this shift, that the element earth naturally moves toward point p. Instead, a shifted version of the principle will hold: "the element earth naturally moves toward point q," where q is three units to the left of p. Formulated in this way and using an active conception of the transformation, the original principle does not say the same thing or make the right predictions after the transformation.

These inferences are, then, more complicated than the previous discussion might have suggested, for whether a law remains the same or says the same thing after a transformation depends on how we formulate the law and how we construe the transformation. This may seem like an overwrought way of making an obvious point. The reason to go through it is to emphasize the fact that, regardless, we are led to the same conclusion about the requisite structure. Formulated in a coordinate-independent way, the above principle does not say the same thing after applying an (active) transformation that shifts objects' locations: this reveals that the principle presupposes a preferred-location structure. Formulated in a coordinate-dependent way, the principle does not retain its form after a (passive) coordinate translation or shift in origin: this, too, reveals that the principle presupposes a preferred-location structure. Either way, we are led to the same structure for the physics. (A structure that in this case is easy to see even without mucking about with transformations, just by considering what the law explicitly refers to; we have come across laws for which the requisite structure is not elicited so easily. Notice that the earlier distinction between explicitly mentioning and implicitly presupposing some structure is evident here: the coordinate-dependent formulation implicitly assumes a distinguished-point structure, underlying the (x_0, y_0, z_0) mentioned in the law, whereas the coordinate-independent formulation explicitly mentions the preferred-point structure. The current case underscores the point that it does not matter, as far as our inferences about structure are concerned, whether the relevant structure is explicitly mentioned or implicitly assumed: either way, we can see that there is a certain structure required by the laws.)

The second thing to note is that there is still something odd about these inferences, especially the kind of inference based on a passive transformation or change in coordinates. In fact, there is something of a tension behind it. For consider the following two thoughts:

(1) Coordinate systems are labeling devices, auxiliary descriptive tools, not inherent in physical systems themselves. Coordinate systems are things we use for convenience, allowing us to describe and predict systems' behavior by means of algebraic equations. But the fact that they are auxiliary labeling devices means that we may use any coordinate system we like to describe any physical system, for any physical theory: any coordinate system is allowed. The choice of coordinate system is completely arbitrary.

(2) The laws may not be invariant under arbitrary changes in coordinate system. In particular (as we just saw), a coordinate-based formulation of the laws may no longer be true after certain coordinate transformations. This means that certain coordinate systems, namely those that don't preserve the truth of the laws, will be disallowed by the theory. The choice of coordinate system is not completely arbitrary.

These two ideas both seem to be true, yet they also seem to be in conflict. How can a coordinate system be a matter of arbitrary choice when that choice can affect the physics? For that matter, how can a change in coordinates, a mere relabeling, in any way affect the physics? Physical principles themselves cannot "say something different" just because we have chosen to label things differently! Physics is independent of our descriptive devices.

You may be led to reject inferences based on passive transformations and coordinate-based formulations of the laws altogether—a not-unfamiliar attitude in foundational discussions. This attitude is reinforced by the thought that coordinate systems, qua descriptive devices, carry no intrinsic physical significance; *ipso facto*, transformations between coordinate systems cannot indicate anything physically significant. John Stachel expresses it this way:

> A coordinate transformation is a relabeling or renaming of the points.... Such a passive transformation does not have any physical significance, since it amounts to nothing more than a mathematical redescription of the same model of a physical theory... [whereas] an active transformation, is of potential physical significance since... it can be used to turn one physical model into another.
>
> (Stachel, 1993, 133)

Something like this seems to be the reason people balk at drawing physical conclusions on the basis of coordinate transformations and coordinate-dependent, even coordinate-based, formulations of the laws. Coordinate systems are conventionally chosen descriptive devices, and mere changes to a conventionally chosen mode of description cannot indicate anything about physical reality. We then resolve the tension between (1) and (2) above by simply jettisoning all theorizing in terms of coordinates.

It is tempting to make the leap from the fact that coordinate systems are labeling devices to the conclusion that any reasoning based on coordinates cannot tell us about physical reality. But the conclusion does not follow. We have seen this by example in the current chapter, as when the fact that Newton's laws retain their form in the coordinates of any classical inertial frame indicates a Galilean spacetime structure, whereas the fact that Maxwell's equations do not so retain their form indicates a different spacetime structure. These facts, though they involve coordinate systems, are not merely about our labeling devices, but flow from the underlying nature of the physical world. We saw something analogous for the Euclidean plane, where the existence of certain coordinate systems, and the invariance in form of the metric under transformations of them, indicates the plane's structure. Coordinate systems are labeling devices that can be chosen for reasons of convenience; nonetheless, various features of coordinate systems, including the behavior of coordinate-based or even coordinate-dependent formulations of the laws under transformations of coordinates, can be physically significant. (Recall the related point from Bell's "How to Teach Special Relativity" discussed in Chapter 2: the mere fact that objects' lengths depend on the chosen reference frame does not entail that any phenomenon involving lengths is wholly physically unreal. More generally, Bell notes that, "the laws of physics in any *one* reference frame account for all physical phenomena, including the observations of moving observers" (1987a, 77; original italics): in other words, even a frame-dependent version of the laws can tell us about physical reality and physically real phenomena.)

It is likewise tempting to dismiss the possibility of drawing physical conclusions on the basis of passive rather than active transformations, as Stachel does. However, we have seen that we can reach the same conclusions about a world's physical structure by reasoning on the basis of either kind of transformation. To consider the form of the laws under various changes in coordinates is to apply a passive transformation to the laws; and for any passive transformation, there is a corresponding active transformation, which yields the same changes to objects' coordinate descriptions. It shouldn't matter whether we consider the active or passive version of a transformation: a passive transformation provides an alternative means of arriving at the same conclusions. We saw this in the case of the principle above, and it can be shown in the case of the Galilean invariance of Newton's laws or the Lorentz invariance of the laws of special relativity as well.[30]

More generally, we can learn about the nature of the physical world from features of the coordinate systems we use to describe it, just as we can learn about the nature of the Euclidean plane from features of the coordinate systems we use to describe it. The example in Chapter 4 will further demonstrate how we can learn about the physical world by reasoning about coordinate systems, all the while acknowledging that coordinates are descriptive devices that invariably possess a degree of arbitrariness.

There is furthermore a way of alleviating the above tension other than by dispensing with coordinate-based reasoning altogether. The first step toward doing so is to recognize the sense in which certain coordinate systems are "allowed" for a given theory. As we started to see in Chapter 2, claims that certain coordinate systems are disallowed can't be quite right as they stand. As a general rule, we can use any coordinate system we like (subject to existence conditions). The sense in which certain coordinate systems are "disallowed" cannot be that of an outright prohibition, despite the connotation of the word. Rather, it has to do with the fact

[30] Brading and Castellani (2007, 1342) note that the following characterizations of passive and active transformations are "equivalent in the sense that they pick out the same set of transformations: (1) Transformations, applied to the independent and dependent variables of the theory in question, that leave the form of the laws unchanged. (2) Transformations that map solutions into solutions." I do think that it is possible for passive and active versions of a transformation to come apart in a way—an example is in North (2008)—but that is a special case.

that something else is not sufficiently preserved or respected by those coordinate systems.

Thus, take Newtonian mechanics and the familiar claim that only inertial reference frames and their coordinate systems are " 'allowable' co-ordinate systems" (Weyl, 1952a, 153), a non-inertial reference frame or coordinate system being the "wrong coordinate system" to use (Feynman et al., 2010, Sec. 12.5). (Again, we may treat coordinate systems and reference frames interchangeably for our purposes. Think of a (non)inertial coordinate system as one that is naturally adapted to the (non)inertial reference frame of an observer.)

Inertial reference frames and coordinate systems are ones in which Newton's laws hold. (A classical inertial frame is often defined to be one in which Newton's laws hold.) Choose a non-inertial reference frame, and the laws no longer seem to hold: things can appear to accelerate even though there is no net external force on them. Consider a particle traveling with uniform velocity along a straight line, and imagine viewing it from the perspective of a reference frame that is turning back and forth. From the perspective of this frame, the particle will appear to be jerking back and forth even though there is no net external force on it, when in reality it is traveling inertially; we have simply described the motion in terms of the coordinates of a non-inertial reference frame. In such a frame, things appear to happen which the theory itself deems physically impossible. Non-inertial frames thus seem prohibited by the theory in that they do not preserve the truth of the laws.

However, we saw in Chapter 2 that this thinking is not quite right. We can use non-inertial reference frames and coordinate systems in Newtonian mechanics, so long as we are careful to use the correspondingly altered equations. The altered equations will contain additional "pseudo force terms," so-called because the things they refer to, if they existed, would not be genuine Newtonian forces, behaving in accordance with the usual Newtonian laws. We must simply remember that there aren't any genuine physical forces corresponding to these terms: their appearance is merely an artifact of using a non-inertial reference frame, not an indicator that the theory is false. (There are some further subtleties about this that we will come to in Chapter 7.)

In other words, certain reference frames or coordinate systems are disallowed in the sense that they don't preserve the usual *form* of the laws, not in that we are prohibited from using them on pain of falsifying the laws. Conversely, the sense in which certain reference frames or coordinate systems are allowed is that "the laws of mechanics take their simplest form" in them (Landau and Lifshitz, 1976, 5), which makes these the preferred ones to use. V. Fock puts it this way: "There exist frames of reference in which the equations of motion have a particularly simple form; in a certain sense these are the most 'natural' frames of reference," which are thereby preferred by the theory, even though "it is always possible, by a mathematical transformation, to pass from a preferential coordinate system to any other arbitrary one" (1964, 15; 158). Or as Weyl puts it, "Two systems of reference are equally admissible if in both of them all... physical laws of nature have the same algebraic expression" (1952b, 129)—which is not to say that other frames can't be used, so long as the algebraic expression is correspondingly altered.

Properly understood, therefore, both (1) and (2) above are true. We can in principle use any coordinate system we like, and this does not contradict the claim that certain ones may nonetheless be privileged or preferred by virtue of yielding a simple or natural statement of the laws. The choice of coordinate system is completely arbitrary in that this is a mere choice of descriptive device. At the same time, the choice of coordinate system is not completely arbitrary in that certain ones yield a natural form of the laws.

This leads to a third note, which is that learning what the preferred coordinate systems (in the above sense) are is an important piece of physical information. For this is indicative of underlying structure, just as the preference of Newton's laws for certain coordinates indicates a Galilean spacetime structure, and the preference of the special relativistic laws for different coordinates indicates a Minkowski spacetime structure—even though we *can* always use other kinds of coordinates. Although the choice of coordinate system is arbitrary in that we can choose to label things however we wish, the fact that certain coordinate systems are preferred or privileged or especially natural in this way is not arbitrary, but stems from the underlying structure. (Fock notes that, "the existence of a preferred set of coordinates... is by no means trivial, but reflects intrinsic properties of space-time" (1964, 374).) Again recall the analogous idea in the mathematical case of the Euclidean plane: the preference for Cartesian coordinates, in the sense that the metric takes a particularly simple, natural form in them, is indicative of underlying structure, even though we can always use other types of coordinates instead.

You may still think it best to eschew all mention of coordinates and coordinate transformations in favor of wholly coordinate-free, geometric formulations of physics. Coordinate-free formulations can make the underlying structure most transparent. In addition, since the laws themselves are about coordinate-independent, physical things, a coordinate-free formulation seems preferable because it more directly reflects the natures of the things the theory is about.[31]

I agree that a direct formulation is in general preferable (for reasons we started to see in Chapter 1 and will see further in Chapter 5), and the more direct formulation will typically not mention coordinates. However, this doesn't change the fact that a formulation that mentions coordinates can still be a good guide to the nature of the physical world. Although coordinate-free approaches have become

[31] Such ideas are expressed throughout Thorne and Blandford (2017), for one, and seem to be assumed by many philosophers of physics nowadays. Discussion is in Wallace (2019), who issues a "call for pluralism" about coordinate-free and coordinate-based approaches to both physics and mathematics.

de rigueur in philosophy of physics, a formulation that mentions coordinates can be equally elucidating. As Wallace says, drawing an analogy between the mathematical and physical cases: "It is generally fine, and often actively useful, to characterise mathematical spaces via classes of preferred coordinatisations of these spaces.... It is, equally, generally fine, and often actively useful, to specify dynamical equations in physics via coordinate expressions" (2019, 135). What we are ultimately after is a physical theory and nature of the world that don't depend on our coordinatizations; but this can be achieved by means of formulations that mention coordinates. The mere mention of coordinates does not mean that the nature of the thing in question is itself coordinate-dependent—as we can see from the fact that the existence of certain kinds of coordinates for the Euclidean plane suffices to characterize its structure. All it means is that, whenever coordinates are mentioned, we will have to be careful to distinguish the features that get at underlying structure from the ones that are artifacts of description.

A related concern, leading to a fourth and final note. An equation's form seems too dependent on the particular type of coordinate system to be physically significant. For that matter, an equation's form seems too vague a notion on which to base any weighty physical conclusions. Surely this is a reason to eschew coordinate-based reasoning in physics, particularly foundations of physics.

As mentioned above, the legitimacy of these inferences is something I am essentially assuming for purposes of this book. This assumption is not without reason, given the ubiquity of these inferences and the apparent plausibility of their conclusions. Open just about any physics book and you are bound to find some physical reasoning based on the form of a theory's equations and how this is affected by various transformations. Consider the following discussion from Leonard Susskind in one of his *Theoretical Minimum* books:

> If you take Maxwell's equations, which contain x's and t's, and plug in the old Galilean rules, $x' = x - vt, t' = t$, you would find that these equations take a different form in the primed coordinates. They don't have the same form as in the unprimed coordinates. However, if you plug the Lorentz transformation into Maxwell's equations, the transformed Maxwell's equations have exactly the same form in the primed coordinates as in the unprimed coordinates.
>
> (Susskind and Friedman, 2017, 61)

As Susskind notes, this kind of reasoning leads to special relativity over classical physics. Bell likewise notes the "exact mathematical fact" that Maxwell's equations and the Lorentz force law, when expressed in terms of new variables related to the old via the Lorentz transformations, "*have exactly the same form as before*" (1987a, 73; original italics). Or consider one of Einstein's own characterizations of his theory of special relativity: "The laws of nature are invariant with respect to Lorentz-transformations (i.e., a law of nature does not change its form if one introduces into it a new inertial system with the help of a Lorentz-transformation on *x*, *y*, *z*, *t*)" (1950, 8). Fock notes more generally that, "An important place in the theory of space and time is taken up by the question of different coordinate systems and the change in the appearance of equations on going from one such system to another" (1964, 4).

Physicists routinely note the sameness in mathematical form, or not, of an equation under various transformations, and they go on to draw substantive physical conclusions from this. They do not take themselves to be discussing mere features of our descriptive schemes. What is more, these discussions generally assume the idea of an equation's form—its "formal appearance" (McCauley, 1997, 247), the "functional form of the law, expressed in terms of the independent and dependent variables" (Brading and Castellani, 2007, 1343)—without further explication. The idea is sufficiently clear to be considered an "exact mathematical fact," in Bell's phrase. Of course, we shouldn't take everything physicists say at face value. But this kind of reasoning is so ubiquitous, and it yields conclusions that are so familiar and seemingly plausible, that it is hard to dismiss it as altogether unjustified. In Goodmanian vein, we may conclude that this mode of reasoning is in general justified.

Since an arbitrary coordinate transformation applied to a coordinate-based equation will in general alter the equation's form, it is significant to discover when this does not happen—to learn when the transformed expression, given in terms of the transformed coordinates, looks exactly the same, the only difference being the primes on the variables. More generally, manipulating the mathematical expression of a physical law is a test for whether the physics itself remains the same under various physical changes. Feynman expresses the idea this way:

> You write the equations with certain letters, then there is a way of changing the letters from x and y to a different x, x', and a different y, y', which is a formula in terms of the old x and y, and the equations look the same, only they have primes all over them. This just means that the other man will see the same thing behaving in his apparatus the same way as I see it in mine.... [T]he laws of physics look the same to him as they do to me. (Feynman, 1965, 88; 90)

Again recall the inference of a Galilean spacetime structure for Newton's laws. It is of physical significance to learn that Newton's equations, when expressed in terms of the coordinates of a given inertial frame, do not change form, in the sense that Feynman is discussing, under the Galilean transformations. As one textbook puts it: "Newton's laws work in my inertial frame, S; my unprimed coordinates obey them. I can then use the Galilean transformation to show that you, in S', will find that your primed ones will obey them as well"; we conclude from this that, "you are going to get the same Newtonian laws" (Shankar, 2014, 200).[32] We draw reasonable physical conclusions from the behavior of the mathematical form of the equations under various transformations.

We are trying to figure out whether a physical theory is sensitive to various alleged features of the world, in which case we infer that these are genuine physical features of the world. One way to check for this is to formulate the laws mathematically and see what happens to the equations under mathematical transformations that correspond to various physical changes. If the equations expressing the laws are unaffected by the relevant mathematical transformations, then it is reasonable to conclude that the laws are indifferent to the corresponding physical changes, and we infer that the world does not possess structure underlying those changes. Investigating the form of the equations under various coordinate transformations is physically relevant. Indeed, it is a test for the very coordinate-independence we want, for it indicates whether the laws "remain the same" regardless of coordinate system, so that the same laws will be confirmed by observers whose reference frames are related by the corresponding coordinate transformations.

Most philosophers of physics nowadays regard coordinate systems as inept and misleading guides to the nature of physical reality, to be dispensed with in foundational discussions. There are good reasons to be wary of coordinate-based representations of physics, but it goes too far in the other direction to eschew all reasoning in terms of coordinates. We mustn't give coordinates outsized significance—coordinates are not intrinsic to the physical things they describe, and not every feature of a coordinate system will directly reflect features of the underlying physical reality—but they can nonetheless tell us about that reality. Coordinate-based reasoning can be an indirect route to the underlying nature of physical (or mathematical) reality, as we have seen in numerous examples so far. The example in Chapter 4 will illustrate this further.

[32] This idea sometimes goes by the name of a "relativity principle," alternately stated as the principle that the form of the equations remains the same in the coordinates of different reference frames, or that observers in different reference frames will observe things to obey the same laws.

4
Classical Mechanics

> The mathematical and philosophical value of the variational method is firmly anchored in this freedom of choice [of coordinates] and the corresponding freedom of arbitrary coordinate transformations.
>
> Cornelius Lanczos (1970, xxv)

4.1 Introduction

We have seen that there is an intuitive way to compare the structures of different objects, spaces, or theories, on the basis of which we can apply familiar epistemic principles such as the minimize-structure rule. We are most familiar with following these rules in our spacetime theorizing, as when we infer that spacetime in special relativity lacks an absolute simultaneity structure.

In this chapter, I further discuss the minimize-structure rule. I am going to suggest that we adhere to a general version of this rule: infer the least structure required by the fundamental laws (both mathematical structure in the formalism and physical structure in the world)—structure in general, not just spacetime structure in particular. As we will see, the general version of the rule has consequences for our inferences about physical space, so that it is not, in the end, all that different from the rule familiarly applied to spacetime structure; either way, the rule guides our inferences from the mathematical structure of a formalism to physical structure in the world. (Indeed, the rule as stated in Chapter 3 was not restricted to spacetime structure, though the examples used to illustrate it were.) The generalized conception of the rule simply invites us to pay attention to any mathematical structure required by the laws, even if that structure is not obviously spatiotemporal.

A corollary of the rule has implications for comparing different theories or formulations: given different formulations of the laws, infer the one that requires the least structure. Choose the formulation that requires the least mathematical structure, and posit the correlatively lesser physical structure in the world. In particular, given two theories that are claimed to be equivalent, we should figure out whether their laws presuppose different amounts of structure. If they do, then the minimize-structure rule will choose between them. By extension, theories or formulations presupposing different amounts of structure are not wholly equivalent.

Physics, Structure, and Reality. Jill North, Oxford University Press (2021). © Jill North.
DOI: 10.1093/oso/9780192894106.003.0004

I am going to illustrate all this here by comparing two formulations of classical mechanics, the Newtonian and Lagrangian formulations, that are standardly regarded as notational variants, different mathematical presentations of a single theory. I argue that their respective laws presuppose mathematical structures that are not only different, but differ in strength or amount. There are also metaphysical differences between the two formulations. (These latter differences I discuss briefly in this chapter, holding off further discussion until Chapter 7.) As a result, contrary to the standard view, these are not wholly equivalent theories, not mere notational variants. They are more like distinct theories, which say different things about the fundamental nature of a classical mechanical world. What is more, since one of them assumes more structure than the other, the minimize-structure rule will choose between them.[1]

In an earlier paper (North, 2009), I discussed the mathematical structures of the Lagrangian and Hamiltonian formulations of classical mechanics. The discussion in this chapter takes off from that paper but goes further. In the earlier paper, I focused on comparing the theories' abstract statespace structures (more on the idea of a theory's statespace below). Here, I aim to highlight a more general notion of "the structure needed for the laws," or a theory's dynamical structure, and what this says about the nature of physical space. Although I discuss the statespace structures of these two theories as well, the focus will be on figuring out what this sort of abstract structure tells us about the nature of the physical world, and how differences in that abstract structure correspond to differences in the physical world. This task is a little more straightforward in the case of Lagrangian versus Newtonian mechanics than Hamiltonian mechanics. A comparison between the Lagrangian and Newtonian formulations will furthermore bring to light what kinds of metaphysical differences there can be even between theories that are standardly claimed to be equivalent, differences that are more marked than those between Hamiltonian and Lagrangian mechanics; the underlying reason being that Lagrangian mechanics—like Hamiltonian mechanics—is a more "coordinate-free" version of classical mechanics than Newtonian mechanics, in ways we will see. (Lagrangian and Hamiltonian mechanics are both exemplars of the "variational method" being referred to in the epigraph.)[2]

Mark Wilson's (2009, 2013, forthcoming) is another dissenting voice on the equivalence of different formulations of classical mechanics, for reasons other than my own. He notes that the various formulations make different, often conflicting assumptions. He further argues that none of them has a clearly delineated content,

[1] One sometimes hears the Newtonian and Lagrangian formulations referred to as "frameworks" rather than theories. Enough assumptions will be made here about the laws and physical ontology that they rise to the level of theories.

[2] A bird's-eye comparison of all three formulations is in North (forthcoming).

for each one has various "descriptive holes," situations that are supposed to fall within the domain of classical mechanics, yet are not obviously treatable by the given formulation. For purposes of this discussion, I will assume that there is a clear-enough domain of content of the theories, but one could question this assumption for the reasons Wilson gives. One can also read into Wilson's work the idea that Newtonian and Lagrangian mechanics are not even empirically equivalent. I will assume for purposes of discussion that the theories are empirically equivalent, as standardly thought, and ask whether there nevertheless remains any significant non-equivalence between them.

Classical mechanics, in any version, is not the fundamental theory of our world (although various aspects are important to candidate fundamental theories, such as Lagrangian mechanics to field theories). Even so, there are some general lessons to be had from thinking about Newtonian and Lagrangian mechanics, taking each one in turn to be a candidate fundamental theory. One is that the task of interpreting a theory, even in the seemingly clear-cut case of classical mechanics, is not so straightforward; in this case, what counts as "the theory of classical mechanics" is itself not so straightforward, let alone what it says about the physical world. Another is a general theme of the book: namely, the importance of a theory's mathematical structure to figuring out what the theory is saying about the nature of the physical world.

A third lesson is that cases of theoretical equivalence in physics are not so easy to come by. What we ordinarily take to be different formulations of a single theory can be more unalike than we had thought. At the same time, cases of genuine underdetermination of theory by evidence are likewise hard to come by, for the theories in question are often not epistemically equivalent. Although I won't try to give an account of theoretical equivalence in this book, I will turn to some general considerations on the topic in Chapter 7. This chapter provides a test case to have in mind for that discussion.

4.2 An overview of the theories

First, let's review some of the main features of Newtonian and Lagrangian mechanics we will need. Many of these features can be extracted from standard textbooks, but it is important to discuss them here, alongside some simple examples. This is because the main points we need are not often explicitly mentioned, and so that those unfamiliar with the theories can follow the discussion. The details highlighted in this section, and illustrated by example in Section 4.3, will be crucial to the argument that follows.

Throughout this discussion, I am going to assume a fundamental ontology of point-particles, or point-masses: point-sized physical objects with the intrinsic feature of mass.[3] Assume that everything in the physical world is made up of point-sized particles, which furthermore move around and interact in a three-dimensional physical space.[4] Think of these as initial physical postulates of the two theories, as I will be discussing them. (More on these sorts of postulates of physical theories in Chapter 7.)

Wilson (2009, 2013, forthcoming) discusses theories of classical mechanics that are not obviously compatible with a point-mass framework, such as the mechanics of rigid bodies and continua. He notes complications and inconsistencies that arise in trying to apply one theory to all these ontologies. He further argues that the principles of various versions of point-particle mechanics are not rigorously derived, so that it is unclear whether any theory of classical mechanics can truly describe systems comprising point-masses. For the sake of this discussion, I am going to assume that Newtonian and Lagrangian mechanics can describe a fundamental point-particle ontology, and I will limit the discussion to such an ontology: my conclusions may not hold given a different fundamental ontology. The complications Wilson notes do imply that the systems in Section 4.3 are strictly speaking "approximate systems," for reasons I will mention. In what follows, I am going to assume that we can make the sorts of assumptions and idealizations that are standardly made in textbook treatments of point-particle mechanics, and still go on to investigate the structure of these theories, just as we can make the assumption that classical mechanics is a consistent fundamental theory and investigate what a world according to it is like, even though we know that it ultimately runs into problems when taken to be a truly fundamental theory. Not everyone will agree with this procedure.

4.2.1 Newtonian mechanics

In the Newtonian mechanics of point-particles, two kinds of quantities, described by means of two different sets of coordinates, specify the fundamental state of a system at a time (this is in addition to the particles' intrinsic features): the positions and velocities (or momenta) of all the particles in the system.

For a single particle moving around in three-dimensional physical space, three coordinates are needed to pick out its position, one along each spatial direction. For two particles, six coordinates are needed to characterize the system's position, three for each particle. In general, $3n$ coordinates are needed to specify the

[3] They also have the intrinsic feature of charge, but strictly speaking this brings in the theory of electromagnetism, for which not all the assumptions made here hold.

[4] As before, for ease of discussion I will generally put things in substantivalist terms. One can translate this talk of space into relationalist-friendly language.

positions of all the particles in a system of n particles. Three more coordinates are needed to specify the velocity of any particle, how the particle's position is changing along each spatial direction. For a system of n particles, then, the total fundamental state is characterized by $6n$ coordinates, three for the position and three for the velocity of each particle in the system. (We will see that constraints on a system reduce the total number of "degrees of freedom" or coordinates needed to completely characterize the state.)

The dynamical laws of a theory describe the different possible motions or histories for a system. In Newtonian mechanics, there is one fundamental dynamical law, Newton's second law, which is in equation form:

$$\Sigma \mathbf{F}_i = m_i \mathbf{a}_i = m_i \ddot{\mathbf{x}}_i. \tag{1}$$

$\Sigma \mathbf{F}_i$ indicates the sum of all the forces—which are vector quantities, written in bold—acting on a given particle labeled by i; m_i is the particle's mass; $\mathbf{a}_i = \ddot{\mathbf{x}}_i$ is the acceleration, the second derivative of position with respect to time, acceleration and position also vector quantities. (A dot over a quantity indicates a derivative with respect to time of that quantity.) The left-hand side of the equation can be written as $\Sigma_{j \neq i} \mathbf{F}_{ij}$, in order to explicitly indicate the sum of all the forces on a given particle (labeled by i) due to all the other particles (labeled by j; this includes particles both internal and external to the system).[5]

Newton's law is a vector equation. (To be pedantic, we may say that the law is represented by this equation, in order to explicitly acknowledge the distinction between the law and an equation that's used to represent it mathematically; for ease of exposition, I often elide the further phrase.) There will be one such equation for each particle in each of the three component directions: three second-order differential equations per particle. These $3n$ second-order equations can be grouped together into one master equation, which describes how the entire system moves through physical space over time. (The i in equation 1 will range from 1 to $3n$, though the masses will come in groups of three, corresponding to the n particles we assumed at the outset.)

Newton's second law gives all the possible histories of any system, for different possible initial states and subject to different forces. Given the actual initial state and the total forces acting on a system (represented by a vector function), integrating the equation (twice) yields a unique solution, or history: the laws are deterministic.[6] Equation 1, in other words, is the fundamental dynamical

[5] Another familiar form of the law is $\Sigma \mathbf{F} = \dot{\mathbf{p}}$, $\mathbf{p}$ the momentum, which is equivalent to the above formulation in terms of structure—it just utilizes different vector quantities. The two versions of the law may nonetheless not be wholly equivalent, for the sorts of reasons discussed in Chapter 7. See Hicks and Schaffer (2017) on other considerations that could suggest they are inequivalent.

[6] I assume that the theory is deterministic, as standardly thought. See Earman (1986); Norton (2008a) for potential counterexamples and Malament (2008); Wilson (2009) for discussion.

equation of the theory, expressing the fundamental dynamical law. It describes the motion of every particle, in any situation. What forces there are will depend on the types of particles involved, and to calculate the forces we will need additional rules, like the law of gravitation. But once given the forces, Newton's second law will predict the behavior of any system of particles.

Note that all the forces are assumed to be governed by Newton's third law: when one particle exerts a force on another, the second simultaneously exerts a force that is equal in magnitude and opposite in direction on the first. This law tells us that forces come in pairs, as a result of interactions between particles. The particular force laws (like the law of gravitation) further indicate that these forces depend only on the intrinsic features of the particles and their spatial separations. More particularly, all forces are central, directed along the line between the particles, and conservative, derivable from a potential. These further restrictions do not strictly follow from the usual "action equals reaction" statement of the third law, but are independent empirical assumptions needed to derive many standard theorems, such as conservation of energy.

(The above amounts to the "strong form" of the third law. The "weak form" says only that all forces are equal and opposite (Spivak, 2010; Wilson, forthcoming). There are questions surrounding the further restrictions that forces be central and conservative, although this is assumed in standard proofs of energy conservation and other central theorems. Newton himself did not restrict forces in this way; yet it is nowadays usually thought that nonconservative forces, such as frictional forces depending on velocity, arise from fundamentally conservative ones. As Feynman asserts: "there are no nonconservative forces!" (Feynman et al., 2010, Sec. 14.4). There are good reasons, empirical and theoretical, to think that Newtonian forces do fundamentally conform to these restrictions.[7] However, one could question these assumptions, in ways that have led a few authors to doubt the equivalence of different formulations of classical mechanics.[8])

According to Newtonian mechanics, particles move around in response to forces that arise by means of interactions with other particles. It is a physical postulate of the theory that the world is made up of point-particles, and that these things behave in various ways as a result of the forces acting between them.

Newtonian mechanics can also be formulated in terms of a statespace. This is a mathematical space in which we represent all the possible fundamental states of a given system or world. The statespace is a kind of possibility space for the theory, where each point represents a different possible fundamental state of the system or world in question. Since $6n$ coordinates are needed to specify the state

[7] See Callender (1995).

[8] Lanczos (1970, 77 n. 1); Gallavotti (1983, Ch. 3, esp. p. 155); Wilson (2009, 2013). Lanczos (1970, Introduction and Ch. 1), and in a different way Hertz (1899), give further reasons to doubt their wholesale equivalence, akin to the kinds of reasons discussed Chapter 7.

of a system, the statespace will have $6n$ dimensions, each point being uniquely picked out by the values of $6n$ coordinates.[9] (You might wonder about the seeming mismatch between $3n$ equations of motion and a $6n$-dimensional statespace. The answer is that the equations are each second-order, so that a total of $6n$ pieces of information, corresponding to the location of a single point in a $6n$-dimensional space, are needed to derive a solution.)

The Newtonian equations, the equations that represent the laws mathematically, can be defined on the theory's statespace. A solution to the equations, given the initial state of a system and all the forces acting on it, then picks out a trajectory through this space, which represents the paths of the particles through physical space. Just as each point in the statespace represents a different possible fundamental state of all the particles in the system, a curve or trajectory through the statespace (parameterized by time[10]) represents a possible history, a possible sequence of particle positions through physical space over time.

4.2.2 Lagrangian mechanics

In Newtonian mechanics, forces cause particles' behavior; fundamental states are characterized by ordinary position and velocity coordinates; and the central dynamical equation is a vector equation. Interestingly, for a theory alleged to be wholly equivalent to Newtonian mechanics, Lagrangian mechanics differs in all these respects.

In the Lagrangian mechanics of point-particles, systems' fundamental states are characterized by so-called generalized coordinates: the generalized positions, q_i, and their first time derivatives, the generalized velocities, $\dot{q}_i$, of all the particles in the system (i ranges from 1 to $3n$ for n particles in three-dimensional space). As in Newtonian mechanics, we need $6n$ coordinates (in addition to the particles' intrinsic features) to completely specify the state of an n-particle system at a time.[11] However, the coordinates are "generalized" in that they need not be, nor need they directly resemble, ordinary spatial positions and velocities. Generalized positions can have dimensions other than that of length, but can be of energy, or length squared, or can even be dimensionless. Any set of independent parameters that suffice to completely characterize the system's state will do, the choice usually made on the basis of the number of degrees of freedom of the system and the topology of the spatial region in which the particles are free to move around. For particles on a

[9] I focus on standard statespace constructions. See Belot (1999, 2000) on others.

[10] Alternatively, time can be included as an additional dimension of the statespace.

[11] Given m constraints on the system, $3n - m$ coordinates are needed to fully characterize the state. In what follows I assume holonomic (essentially, integrable) constraints, for which the relationship between the different coordinates is expressible in terms of a simple algebraic function (there is no rolling or slipping).

two-dimensional surface of a sphere, for example, we might choose the latitude and longitude angles to be the two generalized position coordinates, with generalized velocities the first time derivatives of these angles.[12]

Where Newtonian mechanics predicts a system's behavior given all the forces on it, Lagrangian mechanics requires a scalar function, called the Lagrangian, L, which is typically equal to the kinetic minus potential energy of the system, $L = T - V$. Given a system's Lagrangian and initial state, characterized by the three generalized position and three generalized velocity coordinates of each particle in the system, the following equations, called the Lagrange or Euler-Lagrange equations, yield a unique solution:

$$\frac{d}{dt}\left(\frac{\partial L}{\partial \dot{q}_i}\right) - \frac{\partial L}{\partial q_i} = 0. \tag{2}$$

(These equations can be derived from a variational principle, hence the reference to the "variational method" in the epigraph to this chapter.) As in Newtonian mechanics, here too the motion of a system of n particles in three-dimensional space is given by $3n$ second-order equations, one for each particle in each spatial direction; one for each degree of freedom. Notice, though, the lack of any reference to Newtonian-style forces.

Lagrangian mechanics can also be formulated in terms of a statespace, which will again be $6n$-dimensional. More on this in Section 4.6, but a few things for now. This statespace has the structure of a tangent bundle, which is a space that combines, in a particular way, the $3n$-dimensional space on which we represent the generalized positions—the "configuration space"—with the $3n$-dimensional tangent space at each point. (The tangent spaces are needed to define the generalized velocities, which are tangent to the generalized positions.) Standard labels are Q for the configuration space, the "base space" of the tangent bundle, T_qQ for the tangent spaces—the "fibers," one for each q in Q—and TQ for the entire statespace. (The elements of T_qQ are the different possible generalized velocities for a system in configuration q.) Note that the points in the configuration space represent the different possible configurations of the particles in ordinary space. More generally, the configuration space Q represents the physical space that a system's particles can move around in. Given the freedom in allowable coordinates, this representation needn't occur in an obvious or straightforward way; yet the structure of the system's available physical space will still be coded up in the structure of Q.

[12] There are some mild constraints on generalized coordinates: Lanczos (1970, Sec. 1.2); José and Saletan (1998, Sec. 2.1.2). Wilson (2009) notes that the very idea of generalized coordinates, and the requirements on them, are not as straightforward as usually assumed.

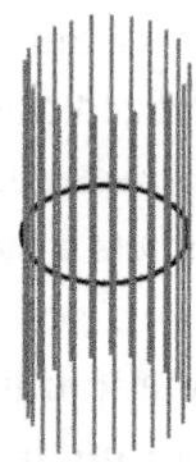

Fig. 4.1 Two-dimensional tangent bundle

To give you a sense of what these spaces look like, here is a depiction of the statespace for a particle moving on a one-dimensional circle: Figure 4.1.[13] This statespace is two-dimensional, each point being picked out by two coordinates $(q, \dot{q})$. The circle represents the different possible values of the generalized position coordinate, q. Each line (called the fiber above q) represents the different possible values of the generalized velocity coordinate, $\dot{q}$, for a given q. (This could be the statespace of a plane pendulum, to give an example discussed below, with the circle representing the different possible values of the angle θ the suspending string makes with respect to the vertical, the lines the different values of $\dot{\theta}$.) To construct this space, we take the configuration base space and join all the tangent spaces together in a smooth and non-overlapping way: this yields the tangent bundle. (The fibers are all tangent to the base space because of how the generalized velocities are defined, and they smoothly connect without crossing because of the differentiability and determinism of the equations.)

In the statespace formulation of the theory, the Lagrangian function, L, assigns a number, a scalar value, to each point in the statespace. (L is a function of q, $\dot{q}$, and the intrinsic features of the particles.[14]) Although this gives the Lagrangian as defined on TQ, we can think of it as coding up information about particles' ordinary spatial features, those that are relevant to their energies, so that it is ultimately about goings-on in three-dimensional space. A solution to the equations, for a given initial state and Lagrangian, yields a trajectory on Q (solutions are given by curves through TQ, which are then projected onto Q), representing the paths of the particles through physical space. Curves through the space depicted in Figure 4.1 (parameterized by time), for instance, represent different possible histories of the particle, possible sequences of fundamental states characterized by $(q, \dot{q})$ over time. Just as in Newtonian mechanics, here too we can represent everything that happens physically by means of a single trajectory through a high-dimensional statespace.

[13] Though this is "just about the only easily visualized nontrivial TQ" (José and Saletan, 1998, 94); with more degrees of freedom, things quickly become difficult to picture.

[14] L can also be a function of t, but set that aside here given the assumption of energy conservation.

4.3 Examples using Newtonian mechanics

Turn to some simple examples to get a feel for how each theory describes physical systems. This will bring to light some key features peculiar to each theory's approach.

Start with Newtonian mechanics. The goal is to find the system's equation of motion, the equation that describes particles' positions in physical space as functions of time, using Newton's second law. For now assume that the equations describe particle motions directly in terms of three-dimensional space. We will return to the statespace construction in the middle of Section 4.6.

To keep things simple, these examples will be of single-particle systems. Some of these involve "constraint forces," forces that constrain the particle to a particular path without doing any work. (Examples of such forces include the normal force that keeps a bead on a wire, or the tension in the suspending string of a pendulum that constrains the bob to move along an arced path.) As a result, at this point already we run into the assumptions and idealizations mentioned earlier. Wilson (2009, 2013) argues that point-mass Newtonian mechanics does not strictly speaking allow for forces of constraint, which are typically velocity-dependent and do not conform to the third law. Indeed, the usual action-reaction requirement of the third law will be relaxed in these examples. It will nonetheless be important, when it comes to a comparison with Lagrangian mechanics, to have these cases in mind, requisite idealizations and all. It is in any case standard fare to treat these systems as governed by the theory.[15]

First consider a single particle of mass m moving along a finite segment of a straight line. Let x be the distance along the line, and let the force on the particle be a function only of x. The equation of motion is $F(x) = m\ddot{x}$ (leaving off the boldface in the case of one-dimensional motion). This is the general equation of motion for a particle moving along a straight line. Plugging in a particular force function then yields the equation for a particle moving under the influence of that type of force.

If the particle undergoes simple harmonic motion, for example, for which $F(x) = -kx$ (k a constant), then the equation of motion is this:

$$-kx = m\ddot{x}. \qquad (3)$$

If the particle falls freely from rest under the force of gravity, then the equation is this:

[15] This leaves room for an alternative route to the conclusion that the theories are not fully equivalent: perhaps only Lagrangian mechanics can consistently handle systems with constraints. As a historical matter, people (including Lagrange himself) noticed that Newton's laws for point-masses seem unable to handle certain systems with constraints, such as a double pendulum. Discussion and references in Smith (2008).

$$g = \ddot{x}. \tag{4}$$

Integrate the equation of motion once for $\dot{x}(t)$ and again for $x(t)$. Plug in the initial state for the particular solution.

Next take a single particle of mass m constrained to move along an arclength of a circle, again with a force that is a function of position along the path: an example is a simple (point-mass) plane pendulum. It seems as though we should be able to find the solution in the same way. However, it turns out to be a bit more difficult. The reason is that Newton's law favors certain kinds of coordinate systems, namely, the rectangular, Cartesian ones. As a result, things are less straightforward when a particle follows a path that is naturally characterized in terms of non-rectangular coordinates.

The favoritism for Cartesian coordinates can be seen from the fact that equation 1 takes a different form in other kinds of coordinates. In polar coordinates, for example, with basis vectors $\mathbf{e}_r$ and $\mathbf{e}_\theta$, Newton's equation is this:

$$\Sigma \mathbf{F} = m\left((\ddot{r} - r\dot{\theta}^2)\mathbf{e}_r + (r\ddot{\theta} + 2\dot{r}\dot{\theta})\mathbf{e}_\theta\right). \tag{5}$$

Or for r constant, as in the case of a pendulum:

$$\Sigma \mathbf{F} = m\left(-r\dot{\theta}^2\mathbf{e}_r + r\ddot{\theta}\mathbf{e}_\theta\right). \tag{6}$$

Unlike equation 1, the term on the right-hand side is not a simple linear function of the acceleration. In particular, it is not the case that $a_i = \ddot{x}_i$ in these coordinates (that is, $a_r \neq \ddot{r}$ and $a_\theta \neq \ddot{\theta}$): additional terms appear, including first derivatives, resulting in a form that differs from the standard one. (Equation 1 effectively assumes a Cartesian coordinate basis, so that it picks up additional terms when we differentiate the position vector with respect to the polar coordinate basis, for we must differentiate the non-constant basis vectors as well.) In this sense, Newton's law prefers or privileges Cartesian coordinates, the type of coordinates in which the central dynamical equation takes its natural or standard mathematical form. (Recall this sense of preferred or privileged or particularly natural coordinates from Chapters 2 and 3.)

Consider the pendulum as an example. Take a vertical plane pendulum, which swings on an arc through the $x - y$ plane, as illustrated in Figure 4.2. Assume the usual idealizations: frictionless, light (massless), rigid string; point-mass bob; negligible air resistance; uniform gravitational field. (Again we run into the question whether the theory can really allow for the force of constraint on the pendulum bob. One reason is that the suspending string, though idealized as massless, might be best construed as a fundamentally extended rigid body. I continue to assume a fundamental point-particle ontology, making the approximations and idealizations necessary to accommodate this assumption, as done in standard textbook treatments.)

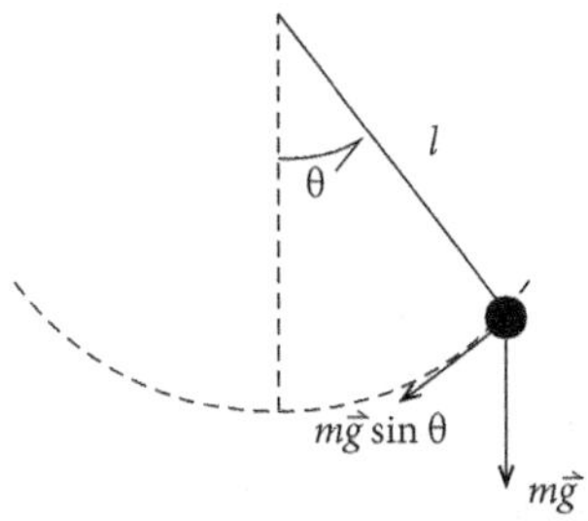

Fig. 4.2 Simple plane pendulum

To apply Newton's second law and solve for the motion of the pendulum bob, choose a rectangular coordinate system, with y in the radial direction and x tangential to the path. This yields two component equations of Newton's law, $F_x = ma_x$ and $F_y = ma_y$, one in each coordinate direction. The component of the gravitational force in the direction of the linear acceleration is $mg\sin\theta$, where θ is the angle the string makes with respect to the vertical. Plugging in this force component yields $F_x = -mg\sin\theta = ma_x$. (The other equation yields $F_y = T - mg\cos\theta = ma_y$, with T the tension in the string. Since $a_y = 0$, we ignore this force and this equation when solving for the equation of motion. T has no component in the direction of nonzero acceleration: it is "merely a constraint force.")

The arclength, s, the distance along the curved path swept out by the pendulum bob, is given by $s = l\theta$, with $\ddot{s} = l\ddot{\theta}$ the acceleration along the path. Plugging this into the x-component equation of Newton's law yields the following equation of motion:

$$-g\sin\theta = l\ddot{\theta}. \tag{7}$$

This equation does not have the same form as the equation of motion for a particle moving on a straight line (equations 3–4). In particular, equation 7 does not involve just a position variable and its time derivatives. The term on the left-hand side is also a function of the sine of that variable: it is "nonlinear in position." (The term "nonlinear" is not used univocally, hence the scare quotes.) The nonlinearity remains if we transform equation 7 back into rectangular coordinates x and y. It is not the use of non-rectangular coordinates per se that yields the different kind of equation. The underlying reason is that the particle follows a path that is naturally characterized in terms of non-rectangular coordinates, along which the acceleration and relevant force component (in the direction of the acceleration) become functions of an angular coordinate.

On its own, Newton's second law does not specify the nature of the forces involved, let alone what kind of function of position they can be. Indeed, the particular force laws (think of Newton's law of gravitation) generally allow for the force terms to be nonlinear functions of position, as is the case for the force term on the left-hand side of equation 7. There is nothing "wrong" with equation 7 as far as Newtonian mechanics in general, or the second law in particular, are concerned. Nonetheless, the pendulum is bringing out the kinds of things that happen when a particle does not travel on a straight path, given the theory's preference for rectangular coordinates and the assumption that forces act along straight lines. For starters, finding a solution becomes less straightforward: we first transform to rectangular coordinates, calculate the various component forces, and plug into the relevant component equations.[16] Things will be different in Lagrangian mechanics.

4.4 Newton's law and Cartesian coordinates

Before considering these examples from the point of view of Lagrangian mechanics, let me say more about why, and in what way, Newtonian mechanics favors certain coordinates; in particular, why "Cartesian coordinates... are central to the point-mass reading of the second law," as Wilson puts it (2013, 71). This is a feature of the theory that tends to go unacknowledged, but it is important. It will be especially important when it comes time to figure out what the theory is saying about the nature of the physical world.

Newton's second law is given by a vector equation, an equation relating vector quantities (in fact, it involves a particular type of vector; more below). Vectors are mathematical objects that transform in a certain way under coordinate transformations. Vectors transform according to a particular rule, which relates a vector's components in one coordinate system to its components in another. (Vectors can be defined as objects that transform in this way: they can be defined by means of the transformation properties of their components.)

Newton's law, qua vector equation, transforms according to this rule. As a result, the law holds component-wise: it says that a net force in a given direction yields an acceleration in that direction. As Feynman puts it, "The general statement of Newton's Second Law for each particle... is true specifically for the *components* of force and momentum in any given direction"; since "any vector equation involves the statement that *each of the components is equal*" (Feynman et al., 2010, Sec. 10.3; 11.6; original italics). And when expressed in terms of components, a vector equation can take a different form depending on the type of coordinates being used, simply because the components of the vector quantities appearing in

[16] We do not always have to resolve forces into their components to solve a problem. Sometimes we work directly with the total force or translate things into the terms of energy. Yet the component-force description more directly reflects what is really going on physically, according to the theory. (See Hicks and Schaffer (2017, Sec. 4.2) on the fundamentality of component forces in Newtonian mechanics.) More on this below and in Chapter 7. (More accurately, a description in terms of natural component forces, in the sense of Creary (1981)—those taken to be genuine physical forces, arising directly from particle interactions rather than the choice of coordinate system—reflect physical reality directly, the latter doing so somewhat less directly.)

the equation can vary depending on the type of coordinates. (A vector equation needn't always take a different form under a change in coordinates. In this case, we will see, it is because the transformation between coordinates is non-linear, hence outside the group of transformations that preserve the equation's form.) Only in Cartesian coordinates does Newton's law take its standard form, $\Sigma F_i = m\ddot{x}_i$, in each component direction. In this way, the theory's focus on vector forces and their components underlies the preference for certain kinds of coordinates.

Since vectors, and tensors more generally, are coordinate-independent, geometric objects, it may seem to follow that Newton's equation cannot privilege any type of coordinate system. Yet although vectors are coordinate-independent objects, they are definable by means of how their components change under changes in coordinates, or basis. Further, there are different types of vectors and tensors, depending on how their components change under different kinds of coordinate transformations. The fact that Newton's equation takes a different form in non-Cartesian coordinates reveals that it is standardly formulated in terms of what are called Cartesian vectors, or Cartesian tensors, which contain an implicit preference for Cartesian coordinates. These objects are vectors, with components that transform in the relevant way under coordinate changes, but the transformations are restricted to those between different Cartesian coordinate systems: Cartesian tensors are invariant under transformations from one rectangular coordinate system to another rectangular coordinate system. When tensors of this type are expressed in terms of components, in other words, we effectively assume a Cartesian (orthonormal) coordinate basis, and implicitly restrict the allowable transformations to those between one such basis and another. A transformation from Cartesian to polar coordinates is not one of these (linear) transformations, which is why the equation changes form under this change in coordinates.[17]

A mathematical object that is definable without reference to coordinates, like a vector, can nonetheless behave differently under different coordinate changes. Relatedly, the fact that an equation is stated in terms of coordinate-independent objects does not entail that it lacks a preference for any kind of coordinates. There is a difference, in short, between existing independently of coordinates and preferring certain coordinates. In a similar vein, physical laws can exhibit a preference for certain coordinates without interrupting the idea that the laws are themselves independent of coordinates. We have come across this before. Recall the special relativistic laws and Lorentz coordinates: the laws are

[17] In other words, Cartesian tensors are defined "without quitting the confines of Euclidean space" (Temple, 2004, 2); more on this aspect below. Cartesian tensors are not usually mentioned in the context of coordinate-free differential geometry and applications to general relativity (they seem more common in engineering and continuum mechanics). See Jeffreys (1931); Temple (2004); Shima and Nakayama (2010, Ch. 18). I am grateful to Mark Wilson for discussion.

coordinate-independent things, but they nonetheless exhibit a preference for certain kinds of coordinates—those in which the laws and spacetime metric take a certain form. We saw this as well for the Euclidean plane and Cartesian coordinates: the plane is a coordinate-independent object, but it nonetheless exhibits a preference for certain kinds of coordinates—those in which the metric takes a certain form. (Compare the discussion of the covariant form of a law at the end of this section.)

There is another, more technical way to see the point about Newton's law. Imagine trying to formulate the law for an intrinsically curved space, like the surface of a sphere, where we cannot use Cartesian coordinates—they don't exist. Velocity will be a vector in such a space, satisfying the usual Leibniz rule. But not acceleration, which has a further derivative that will produce additional Christoffel symbols under a coordinate transformation. In other words, Newton's equation, in its usual form (without the Christoffel symbols, which all vanish in Cartesian coordinates), presupposes that systems move through a space in which we can lay down Cartesian coordinates.[18] Readers familiar with this point may find the discussion of this section belabored, but it is warranted by the fact that the theory's preference for certain coordinates is reasonably unfamiliar. It is rarely stated,[19] and the further conclusion that this is physically significant tends to be objected to, for reasons I began to address in earlier chapters and will continue to do in this one.[20]

In all, Newton's law privileges or prefers Cartesian coordinates. And yet, for systems like the pendulum, these are not the natural coordinates to use. In Cartesian coordinates, the total gravitational force on the pendulum bob always points along a coordinate axis, whereas the acceleration, and the component force in the direction of the acceleration, do not. In polar coordinates, by contrast, both the acceleration and relevant force component remain parallel to a coordinate axis throughout the motion: the quantities of motion vary alongside the basis vectors. Coordinate systems are labeling devices; we can in principle use any coordinate system we like to describe any physical system, with any physical theory. Nonetheless, for the pendulum, there happens to be a particularly natural type of coordinate system, which is not so natural from the perspective of equation 1. It is then odd, or noteworthy, that the usual form of Newton's law assumes Cartesian coordinates.

[18] I am grateful to Sebastian de Haro for this way of putting the point. See Friedman (1983, Ch. 3, eq. 34); Carroll (2004, Ch. 3, eq. 3.56); Ohanian and Ruffini (2013, eq. 7.4) for Newton's law in general coordinates and for curved spaces.

[19] A few exceptions: Friedman (1983, 54–5); Shankar (1994, 80); Wilson (2013, 71). Fetter and Walecka note that the derivation of Lagrange's equations can be seen "as the general transformation of Newton's laws from a cartesian basis" to a more general one (2003, 57).

[20] Another way to reach the conclusion is from the invariance of Newton's laws under Galilean transformations, which presuppose a Euclidean space and corresponding preferred (Cartesian) coordinate systems. The reason for proceeding in the seemingly more roundabout way is to allow for a direct comparison with Lagrangian mechanics, which has not been subjected to the same spatiotemporal scrutiny in the literature.

To say that polar coordinates are natural for the pendulum isn't to say that other coordinates *can't* be used to describe this system. It is to say that polar coordinates are especially well-suited or well-adapted to the motion. Coordinates that are natural in this sense provide "quantitative measures of displacement that are closely correlated with the system's available motions," in Wilson's words; they "relate to the motion in [an] internally 'natural' way" (2013, 71–2). Again, recall the earlier discussions of preferred or privileged or especially natural coordinates: polar coordinates are natural for the pendulum in this sense.

You might still regard the naturalness of polar coordinates as a notational feature without physical significance. We can always transform equation 1 into polar coordinates (as in equations 5 and 6), and use the transformed equation to solve for the pendulum's equation of motion. Alternatively, we can adopt a geometrized formulation of the theory, which, being given in explicitly coordinate-free terms, seems to remove from consideration anything having to do with coordinates. Relatedly, since a mere change in coordinates cannot affect a theory's physical content, it seems to follow that any coordinate-dependent behavior, such as a change in form of the equations under various coordinate transformations, must be physically irrelevant.

We have come across variations of these concerns before. The general worry about coordinate systems is misplaced, since features or behavior that involve coordinates can be indicative of underlying structure, even granting that coordinates are descriptive tools. In the case at hand, it is noteworthy that the usual equation (the one that most directly reflects the theory's usual metaphysics, more on which below) takes a different mathematical form in other kinds of coordinates, which are natural for certain motions; we cannot take $F = m\ddot{x}$ and simply replace x and its time derivatives by θ and its time derivatives. Although coordinate systems are labeling devices that in themselves needn't carry physical significance, a change in form under certain coordinate changes can be physically significant—as is the case for the various inferences about spacetime structure discussed in the previous chapter. And even though the geometrized version seems to suggest that we can reason about the theory entirely without mention of coordinates, so that anything having to do with coordinates must be physically neither here nor there, one might regard the geometrized formulation as a distinct theory altogether,[21] for reasons we will see in Chapter 7. That said, plausibly a similar conclusion concerning the comparison with Lagrangian mechanics follows from the geometrized formulation as the standard one, although I won't discuss this here.[22]

(Although a geometric, coordinate-free version of a theory can be important, a version that mentions coordinates can also be useful, perspicuous, or illuminating, and for that matter might amount to a distinct theory. (Keep in mind that an

[21] As does Earman (1993), for one.

[22] See the references to Friedman's and Malament's discussions in note 34 below.

equation that mentions coordinates can still be coordinate-independent in the sense of retaining its form in different coordinate systems; it can be coordinate-based while not being coordinate-dependent. Lagrangian mechanics will provide an example.) There are several reasons for the focus on equation 1: it is the standard equation, worthwhile investigating for that reason;[23] it is metaphysically perspicuous, most directly reflecting the theory's standard metaphysics; it allows for a particularly clear-cut comparison with Lagrangian mechanics; a different version might not amount to a mere reformulation but a distinct theory.)

Finally, we will see that Lagrangian mechanics eliminates the favoritism for any type of coordinate system, and at the root of this difference lies the centrality of forces to Newtonian mechanics but not to Lagrangian mechanics. In Newtonian mechanics, unlike Lagrangian mechanics, forces fundamentally determine the motion, measured by the acceleration: Newton's laws "say *pay attention to the forces*" (Feynman et al., 2010, Sec. 9.4; original italics). The preference for certain coordinates is no mere mathematical conceit, but part and parcel of the theory's fundamental ontology and dynamical quantities. It is part and parcel of the *physics*.

You might note the distinction between Newton's law, on the one hand, and an equation that is used to represent it, on the other, and go on to conclude that the behavior of the latter cannot indicate anything physically significant. However, as we saw in Chapter 3, we do ordinarily take the behavior of such an equation to matter to the physics, as when we infer a particular spacetime structure by virtue of whether a theory's equations change form under various transformations. Just because Newton's law is not identical to a particular expression of it does not mean that this is a bad inference. Indeed, familiar examples suggest otherwise. (Exactly why inferences based on the form of an equation yield physical insight remains an interesting question. As mentioned in Chapter 3, for the purposes of this discussion I am assuming that this type of inference, ubiquitous in physics, is justified.)

A comparison with the laws of general relativity might seem to indicate that we should not draw these conclusions about Newton's law. In general relativity, the equations expressing the laws can look strange in certain coordinates, yet we don't draw any physical conclusions on the basis of this mathematical feature. The field equations constitute a tensor equation, which, as we know, can look different when formulated in terms of different coordinates, but is itself coordinate-independent. More generally, since we already knew that a vector equation can take a different form when expressed in terms of different coordinates, how can this feature of

[23] Equation 1 is a straightforward mathematical formulation of the law Newton gave us, even though Newton himself did not formulate it this way; that was done by Jacob Hermann in his *Phoronomia* and later by Euler (Smith, 2008). The equation is nonetheless extremely natural—particularly direct—given what Newton says; cf. Truesdell (1968, essays 2–3).

Newton's equation be at all noteworthy, in particular since we do not take it to be noteworthy of Einstein's equations? Answer: it is noteworthy in the face of another formulation that does equally well with respect to the phenomena, yet does not have this feature—the central equation does not take a different form in other kinds of coordinates. There is no similar competitor to Einstein's equations.

That said, it seems as though we can mathematically state Newton's law so that it does not change form under coordinate changes, via the so-called covariant form of the law (Ohanian and Ruffini, 2013, equation 7.4). (This can be arrived at by including the above-mentioned Christoffel symbols, for instance.) This is said to be an alternative mathematical formulation of Newton's law, and it may even seem to be the preferable formulation, by my own lights, since it should require less structure, dispensing with any preferred coordinates. But if that is right, then Newton's law, properly formulated, does not privilege any type of coordinate system, contrary to what I have been arguing.

By using a coordinate-independent version of the tensor calculus, any equation can be given a generally covariant formulation, with a form that is invariant under arbitrary (smooth) coordinate transformations. This is a point of mathematics, and it is true of the equation expressing Newton's law in particular. Since this holds of any equation, for any physical theory, the mere fact that Newton's equation can be cast in generally covariant form is not physically distinctive.[24] What is distinctive is that Newton's law (unlike, say, the laws of general relativity) also has a formulation that is invariant under a smaller group of transformations, which happens to be the standard formulation that is my concern here.[25] And although the minimize-structure principle might seem to tell us to prefer the covariant version to the standard one, it is not obvious that it has the very same content. Putting the equation into covariant form requires the addition of new functions of the variables, which will require physical interpretation. One book notes that the additional terms act like pseudo forces, so that (as in the case of the reformulation of Newton's law for non-inertial frames), "The presence of these extra functions reveals that inertial coordinates retain a special significance" (Ohanian and Ruffini,

[24] This is a reasonably common take on general covariance, albeit not without controversy. Ohanian and Ruffini (2013, Sec. 7.1) is one discussion taking this perspective: "Because equations that are not covariant can be changed into equations that are covariant by the insertion of extra transformation functions, the covariant reformulation of equations is merely a mathematical exercise, without physical significance" (2013, 278). Compare Friedman: "the principle of general covariance has no physical content whatever: it specifies no particular physical theory; rather it merely expresses our commitment to a certain style of formulating physical theories" (1983, 55). See Norton (1993a, 2003) for overviews of the debate surrounding general covariance.

[25] In other words, Newton's law has a formulation with a smaller covariance group, as in equation 1. Wallace notes that one way to conceive of the difference between Newtonian mechanics and general relativity is to compare structure groups (in the ways discussed in Chapter 2), and that, "From this perspective, what is distinctive about general relativity . . . is not that it *can* be formulated on a space with as little (Kleinian) structure as the local diffeomorphism group, but that apparently it *must* be so formulated, whereas other theories have straightforward formulations on much more structured spaces" (2019, 134; original italics).

2013, 278).[26] The covariant equation may not be just an alternative version of the very same law, in other words: it depends on how we interpret the reformulated equation. (We may wish to draw similar conclusions for Newton's law in non-inertial frames; more in Chapter 7.)

Even more importantly, the existence of a covariant form of the law does not eradicate the theory's preference for certain kinds of coordinates (as suggested by Ohanian and Ruffini in the above quotation), just as the formulation of any given law or structure in terms of coordinate-independent mathematical objects does not thereby eradicate all preference for certain coordinates. As Maudlin says, "General covariance does not imply that all coordinate systems are 'equal' or 'the same': there is still a distinction between inertial and non-inertial systems, between accelerated and unaccelerated trajectories" (2011, 213), and likewise between coordinate systems that are better or worse suited to representing these things. Harvey Brown and James Read similarly suggest that there is "nothing intrinsically superior about general covariance; the key issue is which coordinate systems most simplify the form of the equations of the relevant dynamical theory. In the language of Friedman (1983, p. 60), in which coordinates is the 'standard formulation' of the theory obtained?" (forthcoming, 5). Newton's law prefers or privileges certain kinds of coordinates in this sense. That point is unaffected by the existence of a covariant form of the equation.[27]

You might regard the focus on equation 1 as misguided for a different reason. Knox (2014) argues that once we include gravitation and use differential geometry, we see that Newton's law is best formulated differently, in terms of a generally curved spacetime. Regardless, the above applies to the standard formulation and understanding of the theory, which is my focus here. In Chapter 7, we will furthermore see reasons to doubt that Knox's preferred version is merely a reformulation of the very same theory. (That said, as mentioned above, even on the alternative formulation, the primary conclusions for the comparison with Lagrangian mechanics should hold.)

In all, it is intriguing that Newton's law in any way prefers Cartesian coordinates, a preference that's revealed by the change in form of the equation in non-Cartesian coordinates, and by how the theory treats systems naturally characterized in terms

[26] Ohanian and Ruffini (2013, 279) suggest a distinction between the *covariance* of an equation, by which they mean that its form is unchanged under coordinate transformations, and its *invariance*, by which they mean that both the form and the content are unchanged (in that the non-dynamical or "absolute objects," in the sense of Anderson (1967), are unchanged). They note that the covariant formulation of Newton's law is not invariant in this sense. See the discussion in Brading and Castellani (2007), who argue that distinctions such as Ohanian and Ruffini's remain tenable despite ongoing debate about all of these notions.

[27] Wallace (2019, 130) notes that even a generally covariant formulation gives preference to certain types of coordinate systems, viz. the smooth ones, so that it is misleading to characterize it as privileging *no* particular type of coordinate system. He further points out that even if any theory *can* be put in generally covariant form, this is not to say that it *must* be.

of such coordinates. One wonders whether there is a version of the dynamics that eliminates the preference for certain kinds of coordinate systems. Enter Lagrangian mechanics.

4.5 Examples using Lagrangian mechanics

Return to our examples, now using Lagrangian mechanics to find the system's equation of motion.

A single particle moving along a finite segment of a straight line is a system with one degree of freedom, requiring one generalized coordinate to describe the motion. Take the displacement along the line, s, to be the generalized coordinate, with generalized velocity its first time derivative. Insert these into equation 2 to get the general equation for a particle moving along a straight line: $\frac{d}{dt}\left(\frac{\partial L}{\partial \dot{s}}\right) - \frac{\partial L}{\partial s} = 0$.

Plugging in the particular Lagrangian then yields the equation for that kind of motion. For a single particle undergoing simple harmonic motion or free fall, respectively, find L, calculate derivatives, and plug into equation 2 to get:[28]

$$-ks = m\ddot{s} \tag{8}$$

and

$$g = \ddot{s}. \tag{9}$$

Notice we get the same equations of motion we did using Newton's law, with the Lagrangian generalized position coordinate s and its time derivatives replacing the Newtonian position coordinate x and its time derivatives: compare equation 3 with 8 and 4 with 9.

To find the pendulum's equation, we could use rectangular coordinates x and y as we did before. But things are much simpler when we realize that we only need one generalized coordinate, θ, with generalized velocity $\dot{\theta}$, and plug these into equation 2; that is, calculate L in terms of θ and $\dot{\theta}$ and plug this, along with θ and $\dot{\theta}$, into the equation.[29] We get the same equation of motion we did using Newton's law:

$$-g\sin\theta = l\ddot{\theta}. \tag{10}$$

[28] For simple harmonic motion, $L = T - V = \frac{1}{2}m\dot{s}^2 - \frac{1}{2}ks^2$. So $\frac{\partial L}{\partial s} = -ks$, $\frac{\partial L}{\partial \dot{s}} = m\dot{s}$, and $\frac{d}{dt}\left(\frac{\partial L}{\partial \dot{s}}\right) = m\ddot{s}$. For free fall, $L = T - V = \frac{1}{2}m\dot{s}^2 + mgs$, with s the vertical displacement. So $\frac{\partial L}{\partial s} = mg$, $\frac{\partial L}{\partial \dot{s}} = m\dot{s}$, and $\frac{d}{dt}\left(\frac{\partial L}{\partial \dot{s}}\right) = m\ddot{s}$.

[29] The arclength is given by $s = l\theta$, with velocity its first time derivative. Kinetic energy $T = \frac{1}{2}mv^2 = \frac{1}{2}m(l\dot{\theta})^2$; potential energy $V = -mgl\cos\theta$ (setting the zero at the height of the pivot point with $\theta = \frac{\pi}{2}$); so $L = T - V = \frac{1}{2}m(l\dot{\theta})^2 + mgl\cos\theta$. Calculate derivatives, $\frac{\partial L}{\partial q} = \frac{\partial L}{\partial \theta} = -mgl\sin\theta$ and $\frac{\partial L}{\partial \dot{q}} = \frac{\partial L}{\partial \dot{\theta}} = ml^2\dot{\theta}$; so that $\frac{d}{dt}\left(\frac{\partial L}{\partial \dot{\theta}}\right) = ml^2\ddot{\theta}$.

As in Newtonian mechanics, the pendulum's equation looks different from the one for motion along a straight line (equations 8 and 9): the former does not just contain a position coordinate and its time derivatives, it also contains the sine of that coordinate. However, the dynamical law from which we derive this equation does not change form in these coordinates. Unlike in Newtonian mechanics, we treat θ and $\dot{\theta}$ as though they are ordinary position and velocity coordinates, and plug them directly into equation 2 to get equation 10.

In more detail. Take the Lagrangian function, $L(q, \dot{q})$, in terms of an original set of (unprimed) generalized coordinates. L can be rewritten in terms of another set of (primed) coordinates by substituting the functional expressions for the q_i and $\dot{q}_i$ in terms of the q'_i and $\dot{q}'_i$. This yields a transformed Lagrangian function of the primed coordinates, $L'(q', \dot{q}') \equiv L\left(q(q'), \dot{q}(q', \dot{q}')\right) \equiv L(q, \dot{q})$. It follows from equation 2 that $\frac{d}{dt}\left(\frac{\partial L'}{\partial \dot{q}'_i}\right) - \frac{\partial L'}{\partial q'_i} = 0$—the same equation, just with primed coordinates replacing the original ones. The form of equation 2 is the same when expressed in terms of any set of generalized coordinates, the only difference being the primes appearing on the variables. (L itself can be a different function of the new coordinates, but its numerical value at any point remains the same. This is what it takes for the Lagrangian, a scalar function, or zero-rank tensor, to be the same regardless of coordinates.)

In Newtonian mechanics, we cannot simply take $F = m\ddot{x}$ and replace x and its derivatives by θ and its derivatives. Newton's law, properly transformed, still holds in other coordinate systems; it will yield the right equation of motion. But the equation representing the law changes form in non-rectangular coordinates. If we were to try to retain the original form of the law in polar coordinates, we would no longer get the right equation of motion, the one that yields the right predictions. Not so the Lagrangian equation, for which we can choose any kind of generalized coordinate and plug into the very same form of equation: just substitute s and $\dot{s}$, or θ and $\dot{\theta}$, for q and $\dot{q}$ in equation 2.

In this way, as one textbook puts it, "Lagrange's equations, unlike Newton's, take the same form in any coordinate system" (Taylor, 2005, 237). The Euler–Lagrange equations are "form invariant under an arbitrary change of coordinates. This form invariance must be contrasted with the Newtonian equation . . . , which presumes that the x_i are Cartesian. If one trades the x_i for another non-Cartesian set of q_i, [the equation] will have a different form" (Shankar, 1994, 80). Sometimes this is put in terms of a "point transformation," which is essentially a coordinate transformation.[30] Thus, not all point transformations preserve the form of Newton's equation, whereas the Lagrangian equations "remain invariant with respect to arbitrary point transformations of . . . coordinates" (Lanczos, 1970, 195).

[30] So-called because the points of the two reference frames or coordinate systems are in one–one correspondence (McCauley, 1997, 53–4).

(You may wonder how much of this discussion hinges on one's views on the metaphysics of laws of nature. It largely holds regardless. Some, like the Humean, might say that the law for a lone pendulum world is given by equation 10 rather than the full Euler–Lagrange equations or Newton's equation. The typical treatment in textbooks, which I appropriate here, is that equation 1 or 2 is the law, from which we derive the equation of motion for a particular system or world, given the initial conditions and relevant force or energy function. If you disagree, then you may consider this discussion as applying to all the laws, some of which you might then regard as "meta-laws."[31])

As we saw in the examples above, using either theory, we find the same equation of motion for any system, and hence the same set of solutions and the same empirical predictions. This is generally the case (or so I am assuming for the sake of this discussion). Textbooks will demonstrate the inter-derivability of the two formulations' laws, and conclude that the theories are fully equivalent. One book concludes that Lagrange's formulation of mechanics "is completely equivalent to Newton's" (Hand and Finch, 1998, 23). Another says that, "the physical content of Lagrange's equations is the same as that of Newton's," the former being "in fact a restatement of Newton's laws" (José and Saletan, 1998, 48; 65). Another says that, "Lagrangian dynamics does not constitute a *new* theory in any sense of the word," for Newton's and Lagrange's equations "have been shown to be entirely equivalent" (Marion and Thornton, 1995, 262). In short, Lagrangian and Newtonian mechanics are "essentially the same theory" (Shankar, 1994, 75). This is the usual view: Lagrangian and Newtonian mechanics are wholly equivalent theories, mere notational variants, differing at most in calculational ease.[32]

4.6 Cross-structural comparison

We have to be careful when claiming an equivalence between two things, however. For two things can be equivalent in certain respects while being inequivalent in others. The question for us is whether, the above-mentioned equivalence between Newtonian and Lagrangian mechanics notwithstanding, there are any other significant respects in which they fail to be equivalent.

At the end of Chapter 2, we saw that mathematical objects or spaces can be characterized in terms of the mappings that preserve their structure. Two mathematical objects or spaces are then said to be equivalent when there is the relevant structure-preserving mapping between them. In order for two sets to be

[31] Thanks to Gordon Belot for this point.

[32] Further intimations of their equivalence can be found in Feynman (1965, Ch. 2); Symon (1971, 3); McCauley (1997, Ch. 2); Talman (2000); Goldstein et al. (2004); Baez and Wise (2005); Taylor (2005); Susskind and Hrabovsky (2013).

structurally equivalent, for example, it suffices that they have the same cardinality; then there will be a bijection between them, a mapping that preserves their structure as sets. When it comes to objects that have additional structure beyond that of a set, there will be further requirements on the mapping used to test for their equivalence. For two groups to be structurally equivalent, for example, the relevant mapping must put their elements in one–one correspondence while also preserving the identity and group operations, thereby preserving their structure as groups. In other words: whether two mathematical objects or spaces are similar in some way or with respect to some structure depends on the structure preserved by the mapping used to test for their equivalence; the objects are then similar, or not, with respect to that structure. As a result, two mathematical objects that are equivalent in some ways or with respect to some structure can nonetheless differ in other ways, with respect to other structure.

In physics, too, we can compare two objects or spaces, even entire theories, by means of the relevant structure-preserving mapping, with the result that the things being compared can be equivalent with respect to some structure while differing in other structure. For example, we saw that Galilean and Newtonian spacetime share some structure; yet since the latter has an additional absolute space structure, we may conclude that it has more structure overall. (This is one reason there needn't be a definitive verdict, for any two structured objects, that one of them has more structure than the other. These comparisons depend on the type of structure being considered, and things can differ with respect to different sorts of structure, in ways that can result in potentially incomparable structures (recall the case of affine versus conformal structure from Section 2.4), or that can render a particular judgment, though plausible, not absolutely clear-cut (recall the example of a topological space with a preferred point versus a metric space from Section 2.4, or the case of Lorentzian versus Einsteinian conceptions of special relativistic spacetime from Section 3.3). The examples discussed in this book reveal that we nonetheless fruitfully compare many things in this way. Indeed, science would be paralyzed if we could only rely on theoretical criteria guaranteed to yield definitive verdicts that one thing is preferable to another in any given case—just think of our reliance on simplicity judgments.)

With that in mind, let's now compare structures for Newtonian and Lagrangian mechanics, in the same ways we generally compare structures in physics and mathematics. This will reveal that, even though Newtonian and Lagrangian mechanics may be equivalent in the ways ordinarily claimed, they are not equivalent in another, important way—their dynamical structure, the mathematical structure presupposed by their respective dynamical laws. Taking this mathematical structure seriously, in the same way we do for our familiar inferences about spacetime structure, this means that the theories differ in the physical structure of the worlds they describe. (In Section 4.7 I elaborate on the physical differences mentioned only briefly here.)

Recall the inferences discussed in Chapter 3. On the basis of the transformations that do, and those that do not, preserve the form of the laws, we infer that a certain spatial, temporal, or spatiotemporal structure underlies those laws. If the laws retain their form under an inversion of the time coordinate, for example (as do Newton's laws), we infer that the temporal structure according to the theory is symmetric; whereas if the laws do not retain their form under this transformation (as in a law of wavefunction collapse), we infer that there is a temporal orientation, a structural difference between the two temporal directions. If the laws retain their form under spatial translations (as in the case of Newton's laws), we infer that space is homogeneous. If the laws do not retain their form under spatial translations, we infer that there is a structural difference among different spatial locations: not all spatial points are alike; certain ones are preferred by the theory. This is the kind of thing we infer from the principles of Aristotle's physics.[33]

Recall the case of special relativity in particular. The laws remain the same regardless of choice of simultaneity frame. The dynamical equations, when expressed in a different Lorentz frame with coordinates (x', y', z', t'), always have the same mathematical form they did in the original frame with coordinates (x, y, z, t), the only difference being the primes on the variables. The equations have the same form when expressed in terms of the new frame's coordinates as they did in the old. This reveals that the laws do not distinguish or recognize differences among different Lorentz frames—they say the same thing regardless. This, in turn, means that they do not require or presuppose the mathematical structure that would underlie a distinguished or preferred such frame. Since that would be additional mathematical structure, and since the laws do not require or presuppose it in their formulation—since this structure isn't needed to support the laws—we infer, with the minimize-structure rule, that the world lacks the corresponding physical structure. We infer that according to special relativity, taking this to be the fundamental physical theory, there is no absolute simultaneity structure in the world.

Apply this type of reasoning to the case at hand. In Lagrangian mechanics, the laws remain the same regardless of choice of coordinates. The dynamical equations, when expressed in a different coordinate system, always have the same mathematical form they did in the original coordinate system, the only difference being the primes on the variables. The equations have the same form when expressed in terms of the new coordinate system as they did in the old. This reveals that the laws do not distinguish or recognize differences among different coordinate systems—they say the same thing regardless. This, in turn, means that they do not require or presuppose the mathematical structure that would underlie a distinguished or preferred type of coordinate system. Since that would be

[33] I say "this is the kind of thing we infer" because performing a spatial translation in the case of the principles of Aristotle's physics is complicated by the fact that the universe is spherical and spatially finite. But that is the gist of what we infer from Aristotle's principles, as can be seen more clearly in the case of the relevantly similar principle discussed in Section 3.5.

additional mathematical structure, and since the laws do not presuppose or require it in their formulation—since this structure isn't needed to support the laws—we infer, with the minimize-structure rule, that the world lacks the corresponding physical structure. We infer that according to Lagrangian mechanics, taking this to be the fundamental physical theory, there is no physical structure in the world underlying a preferred type of coordinate system.

In Newtonian mechanics, the laws do not remain the same regardless of choice of coordinates. The central dynamical equation, when expressed in a different coordinate system, does not always have the same mathematical form it did in the original coordinate system: the equation needn't have the same form when expressed in terms of the new coordinate system as it did in the old. (There can be more of a difference than the mere appearance of primes on the variables, as in the case of a transformation to polar coordinates (equations 5 and 6).) This reveals that the law does distinguish or recognize differences among different coordinate systems—it does not say the same thing regardless. This, in turn, means that the law requires or presupposes the mathematical structure that underlies the preferred type of coordinate system. Since the law presupposes this in its formulation, we infer, in accordance with the principle to ascribe to the world the structure needed to support the fundamental laws, that the world has the corresponding physical structure. We infer that according to Newtonian mechanics, taking this to be the fundamental physical theory, there is physical structure in the world underlying the preferred type of coordinate system.

Notice the parallelism among these inferences. The invariance in form of the Lagrangian equations under coordinate transformations reveals that no type of coordinate system is preferred by this theory, in the same way the laws of special relativity do not prefer any simultaneity frame and time reversal invariant laws do not prefer either temporal direction. The Newtonian equations do change form under certain coordinate changes, revealing that there is a preferred kind of coordinate system for this theory, just as there is a preferred origin for Aristotle's physics. In particular, Newtonian mechanics prefers the Cartesian coordinate systems, in which the dynamical law retains its natural, standard form. Hence, the theory presupposes the structure that underlies the preference for this type of coordinate system—which is a Euclidean spatial structure—in the same way that a spatial structure with a preferred location underlies the privileged origin of Aristotle's laws.[34]

[34] Arnold (1989, 1), for one, notes that, "Newtonian mechanics studies the motion of a system of point masses in three-dimensional euclidean space. The basic ideas and theorems ... are invariant with respect to the six-dimensional group of euclidean motions of this space." Newton himself, of course, simply assumed a Euclidean space. Various textbook discussions of Newtonian mechanics (such as José and Saletan (1998, Ch. 1); Thorne and Blandford (2017, Ch. 1)) build this structure in from the beginning, as do Friedman's (1983, Ch. 3) geometric formulations of Newton's laws. Malament shows that, given the Cartan-style geometric formulation of Newtonian gravitation, and assuming it to be an appropriate classical limit of general relativity, there is "an interesting sense in which Newtonian

In other words: there is a difference in dynamical structure, the structure required by the theories' dynamical laws. Newtonian and Lagrangian mechanics require or presuppose different dynamical structures. Not only that, but the Newtonian laws presuppose more such structure than the Lagrangian laws, for they presuppose the additional structure needed to pick out a preferred type of coordinate system—just as Lorentz's physics presupposes additional structure compared to Einstein's (the structure needed to pick out a preferred rest frame); Newton's physics set in Newtonian spacetime requires additional structure compared to when it is set in Galilean spacetime (the structure needed to pick out an absolute rest frame); and Aristotle's physics presupposes additional structure compared to a theory on which space is homogeneous (the structure needed to pick out a preferred spatial location). (Note that it is not the use of generalized coordinates per se that signals the different, lesser structure for Lagrangian mechanics, just as the mere use of non-Cartesian coordinates for the Euclidean plane does not thereby signal a non-Euclidean structure. It is the theory's failure to privilege Cartesian coordinates, in the sense discussed above, plus the minimize-structure rule that indicates this.)

Newton's law privileges certain coordinate systems in the same way the Euclidean plane privileges certain coordinate systems. None of this means that we can't use other coordinate systems for Newtonian mechanics. But it does mean that certain ones will be particularly well-adapted to the theory and its underlying structure—just as Cartesian coordinates are particularly well-adapted to the structure of the Euclidean plane, even though we can always use other kinds of coordinates on the plane. More generally, remember, although coordinate systems are descriptive devices that can be chosen for reasons of convenience, the extent to which certain coordinates are particularly natural, the fact that they mesh with some feature or structure in an especially natural way, is not just a matter of convenience, but flows from the underlying structure. Hence, we can learn about what that structure is by considering what those preferred coordinate systems are: again, recall the case of the Euclidean plane. Compare Arntzenius' remark that, plausibly,

> the existence of... coordinate systems relative to which the laws take a certain simple form, is due to the fact that there is some fundamental structure of reality which features in the laws, so that the laws take a particularly simple form relative to coordinate systems that are adapted to that structure.
>
> (Arntzenius, 2012, 91)[35]

physics *must* posit that space... is Euclidean" (1986, 182; original italics). I argue for this conclusion in a different way, on the basis of the standard version of the law, with its standard metaphysics, and considerations that are intrinsic to classical mechanics. Indeed, as noted above, on some views the geometrized version amounts to a distinct theory altogether; more on which in Chapter 7.

[35] Arntzenius' aim in this passage is to make a different but related point. He is arguing against taking the existence of certain coordinate systems to be a "rock bottom fact about reality," and therefore against taking the formulation of a law in terms of the existence of such coordinate systems to be "a plausible fundamental law." More on this idea in Chapter 5.

A preference for certain coordinates, in the sense that the laws take a simple or natural form in them, is indicative of, it is evidence for, underlying structure.

In a way, this difference between the theories should not be entirely surprising. Lagrangian mechanics describes systems' behavior in terms of a scalar energy function. Newtonian mechanics describes things in terms of forces, which are vector quantities. As one book puts it, "Newton's theory bases everything on two fundamental vectors: 'momentum' [or acceleration] and 'force' "; whereas Lagrangian mechanics "bases the entire study of equilibrium and motion on two fundamental scalar quantities, the 'kinetic energy' and the 'work function', the latter frequently replaceable by the 'potential energy' " (Lanczos, 1970, xxiv; xxi). Although vectors are coordinate-independent objects, their components change with the coordinate system. Scalars are more coordinate-independent than that, being completely invariant to coordinate changes, not even "altering component-wise."[36] The centrality of (vector) forces to Newtonian mechanics, as opposed to the (scalar) energy functions of Lagrangian mechanics, underlies the difference between the theories, their relative freedom in coordinates in particular. In Lagrangian mechanics, there is no requirement that generalized coordinates must be capable of being grouped into the components of a vector—for a spherical pendulum, for example, generalized coordinates might be the longitude and latitude angles, which are not the components of a vector—as required of Newtonian forces and accelerations.[37] (Consider too the derivation of Lagrange's equations from a least (stationary) action principle: the condition for a curve to be an extremal of a function is independent of coordinate system (Arnold, 1989, 59). This is often given as the underlying reason for the equations' holding in any coordinate system.[38])

The structural differences between the theories thus go hand in hand with particular metaphysical—or simply physical—differences between them. According to Newtonian mechanics, the world is fundamentally made up of particles that move around and interact in response to the forces on them. According to Lagrangian mechanics, particles move around and interact as a result of their energies. Although energy and force functions are mathematically inter-derivable in ways that physics books will show (albeit under certain contestable assumptions: note 8), these theories paint different pictures of the world, centered on differ-

[36] You might want to say that vector components are covariant, varying with the coordinate system, whereas scalars are invariant. I refrain from putting it this way since these terms remain controversial, with meanings that differ among different authors.

[37] Cf. Goldstein et al. (2004, 14).

[38] For instance by Mac Lane (1986, 281–2); Butterfield (2004); Taylor (2005, Sec. 7.1). As Shankar (1994, 80) notes, there are two ways of verifying the coordinate-independence of Lagrange's equations, either "by brute force," i.e. choosing another set of variables and showing that an equation of the same form results, or by deriving the equations from the minimum action principle and noting that the derivation nowhere assumes Cartesian coordinates.

ent dynamical quantities, yielding different explanations of particles' behavior.[39] As Allori (2015a, 2019) argues, we cannot investigate the invariances—more generally, I would say, the structure—of a theory wholly independently of its metaphysics.[40] I will return to these kinds of metaphysical differences between theories, including further subtleties about what I said here, in Chapter 7.

You may wonder about the idea that Lagrangian mechanics construes everything in terms of particles and their energies in three-dimensional space. When the Lagrangian function is stated in terms of coordinates that are not ordinary position and velocity coordinates, this metaphysical picture is less easy to discern. However, I am taking those things to be among the basic physical posits of the theory. These are natural assumptions, given the standard dynamical equations, and this picture continues to hold even when a particular choice of coordinates makes it less obvious.[41] At the same time, we may take this very freedom in coordinates—the fact that any coordinates, even ones that do not appear to correspond to ordinary positions and velocities, are equally capable of capturing the dynamical facts—as evidence of the theory's underlying structure.

Recall from Section 3.3 that some philosophers have recently argued that Newtonian physics requires less than a Galilean spacetime structure. We can now see why the minimize-structure rule on its own does not support such a conclusion: it depends on how we construe this physics. The alternative spacetime structure defended by Simon Saunders (2013), for example, does not support a distinction between inertial and linearly accelerated motion, nor the concomitant distinction between forced and unforced motion, as required of the usual conception of Newtonian forces.[42] For Knox (2014), the correct structure builds the effects of gravity into the spacetime. This, too, contravenes the theory's usual metaphysics, according to which gravity is a force of interaction. My aim in this chapter is to compare a standard or natural conception of Newtonian mechanics with Lagrangian mechanics (though again, the main conclusion about their comparative structure plausibly holds regardless). More on theories' "metaphysical aspects" in Chapter 7.

We reach a similar conclusion about the theories' dynamical structures—both that they differ in structure, and that one of them has more such structure—by comparing statespaces. A theory's laws can be defined on a statespace with a certain

[39] This runs counter to the usual view that, "it's the Newtonian formulation, not the Lagrangian one, that explicitly represents the ontology" (Coffey, 2014, 831), with Lagrangian mechanics being simply a reformulation of Newtonian mechanics. More on this in Chapter 7.

[40] Two other instances of the general idea: Callender (1995) notes that whether a theory is time reversal invariant depends on its fundamental ontology, and Maudlin (2002) says something similar for the question of whether a theory is deterministic.

[41] This basic picture is outlined by Lanczos (1970), for one.

[42] I thank Katherine Brading for this point. Compare Weatherall (forthcoming, 17): in accepting that spacetime structure, "one would need to revise both the conceptual and mathematical foundations of Newtonian physics."

structure. The mathematical structure needed for the laws will then be reflected in the structure of the statespace. We have just seen that the Newtonian laws require more structure than the Lagrangian laws. This means that the Newtonian statespace must have more structure, as we can also see by directly considering the structures of these spaces.

As mentioned earlier, the Lagrangian statespace has the structure of a tangent bundle.[43] A tangent bundle is a particular kind of vector bundle, which is in turn a special kind of fiber bundle. A fiber bundle is a manifold that can be divided up into one space, the base space, plus different fibers, one attaching to each point in the base space. In a vector bundle in particular, the fibers are all vector spaces, and in a tangent bundle more particularly, the fibers are all tangent to the base space. (Recall Figure 4.1.) Note that any tangent bundle will locally "look" like a product space—locally, it is everywhere a product space of the base space and fiber—yet globally, the topology can be different, or "nontrivial on the whole."[44] Think of a Möbius strip, which everywhere has a locally trivial structure—it looks locally everywhere like an ordinary product space—but it is "twisted" globally.

The Newtonian statespace can also be seen as a fiber bundle, but of a certain kind. Since Newton's equations assume that any motion can be described in Cartesian coordinates, the statespace on which these equations are defined, in particular the configuration space that represents the physical space the system moves around in, must admit of such coordinates. This means that the base space is an intrinsically flat ($3n$-dimensional) Euclidean space, with a Euclidean metric—the kind of space on which we can lay down Cartesian coordinates—and that the statespace as a whole is the trivial bundle, with a global product topology. You won't find this claim in physics texts, which don't generally present the theory in terms of a statespace. But it is plausible. A theory's statespace is a mathematical space on which we define the equations that describe any system's history. If those equations presuppose some mathematical structure, then that structure must be possessed by the space on which they are defined. (The statespace formulation of Newtonian particle mechanics in Arntzenius (2012, Sec. 3.4), for one, notably assumes a Euclidean metric on a flat $3n$-dimensional configuration space.)

The configuration manifold (the base space) of the Lagrangian statespace also has a natural metric. But in this case it is a Riemannian metric, a generalization of the Euclidean metric, applicable to arbitrary curved spaces and coordinate systems. The Riemannian metric is required in order to define the kinetic energy term of the Lagrangian, which is a function of $\dot{q}^2 = \dot{q} \cdot \dot{q}$, with $\cdot$ the inner product.[45]

[43] See José and Saletan (1998, Sec. 2.4) for more on this.

[44] I.e., as a manifold, TM is not always diffeomorphic to the product manifold $M \times \mathbb{R}^n$.

[45] The (smoothly varying) inner product naturally induces a norm, which induces a metric (Spivak, 2005, Ch. 9). Although the coordinate-free version of the equations (José and Saletan, 1998, eq. 3.87) does not explicitly mention the metric, the above suggests that it is nonetheless implicitly required. The development of the coordinate-free version in Crampin and Pirani (1986, Sec. 13.8), for one, suggests this.

(The Lagrangian $L : TQ \to \mathbb{R}$ typically has the form $L = T - V = \frac{1}{2}\langle\, , \rangle - V$ (with kinetic energy T, potential energy V, and setting particle mass $= 1$) for a Riemannian metric $\langle\, , \rangle$ on Q, also expressible as $L = \frac{1}{2}\Sigma_{i,j=1}^{n} g_{ij}\dot{q}^i\dot{q}^j - V$, with g_{ij} the Riemannian metric.[46]) This metric gives distances between nearby points, the lengths of curves, and the geodesics, but it differs from the metric of Newtonian mechanics in that the configuration space needn't be intrinsically flat. Typically, it won't be: consider as simple a system as a spherical pendulum, with configuration space the two-sphere. Unlike in Newtonian mechanics, where forces constrain a system to within a certain spatial region, the Lagrangian configuration space itself directly reflects the space the particles can move around in, the system's "arena of motion" (José and Saletan, 1998, 48).

(As in the case of Newtonian mechanics, here too I assume conservative systems, for which the Lagrangian is regular and can take the form of a homogeneous function of the $\dot{q}^2$'s, what V. I. Arnold calls "natural" Lagrangian systems (1989, Sec. 4.19).[47] The thought is that, at bottom, any classical mechanical system or world will be like this, analogous to the thought that, at bottom, there are no non-conservative forces in Newtonian mechanics. There are more general versions of the theory that do not require this assumption.[48] I suspect that my conclusions for the comparative structure of Newtonian and Lagrangian mechanics will hold even in the absence of this restriction, since removing the condition on L should only yield a theory with even less structure.[49] Regardless, the theory must be able to handle systems with regular Lagrangians, and so it must possess the structure needed to do so. Incidentally, the usual route to proving the equivalence of Lagrangian and Newtonian mechanics assumes that L has the standard form, so that a non-equivalence even given this assumption would be especially noteworthy.[50])

[46] See Lanczos (1970, Sec. 1.5); Mac Lane (1986, 286–7); Arnold (1989, 84); Marsden (1992, Sec. 2.4); Szekeres (2004, 469); Spivak (2010, 476).

[47] In particular, coordinate transformations are restricted to time-independent ones, so that we ignore the non-invariance of the Lagrangian under velocity boosts (Butterfield, 2004, 59–60). José and Saletan (1998, Ch. 2) discuss conditions for the Lagrangian to have this form.

[48] Canonical examples in which L does not have this form come from outside the domain of point-particle mechanics assumed here, such as electromagnetism or special relativity. See Shankar (1994, 79); José and Saletan (1998, Sec. 2.2.4); Goldstein et al. (2004, Sec. 7.9) for examples.

[49] For the kinds of reasons in Arnold (1989, Chs. 3–4).

[50] Curiel says that, "A Riemannian metric on a tangent bundle by itself is neither necessary nor sufficient for a Lagrangian representation of a system" (2014, 292, n. 30). Against this, it is plausibly required, for the above reasons. (I do not claim that it is sufficient; the Lagrangian is also needed to specify a system, for instance.) The fact that it is mentioned in numerous discussions (in particular as being needed for the kinetic energy term) bolsters the claim that it is required by the standard conception of the theory: Lanczos (1970, esp. Secs. 1.5, 5.7); Synge and Schild (1978, 168–9); Mac Lane (1986, 286–7); Arnold (1989, Ch. 4); McCauley (1997, Ch. 10); Arnold and Givental (2001, 43); Belot (2003, 403, n. 22); Butterfield (2004, 42); Szekeres (2004, Sec. 16.5); Burns and Gidea (2005, 114); Spivak (2010, part 3). Curiel says that a different structure is required, one that he argues is equivalent in a certain sense to the structure of Newtonian mechanics. His conclusion is based on a different conception of the theories; for instance, he rejects the ideas that in Lagrangian mechanics generalized positions are fundamental and generalized velocities are defined in terms of them, and that configuration space fundamentally represents particle positions. See Barrett (2019) on

The theories' statespaces share some structure. Both are fiber bundles of the same dimensionality, with structure picking out a class of geodesics. Both have a local product topology. There will even be a vector bundle isomorphism between them (a local, linear, invertible fiber-wise map), demonstrating their equivalence at the level of vector bundle structure. The statespaces share some structure, but they differ in other structure. The Lagrangian statespace has a general structure of which the Newtonian statespace is a special kind. The flat structure and Euclidean metric of the Newtonian statespace is a special case of the arbitrary curved structure and Riemannian metric of the Lagrangian statespace. (Compare Arnold: "A newtonian . . . system is a particular case of a lagrangian system (the configuration space in this case is euclidean)" (1989, 53); though note that I ultimately reject the idea that Newtonian systems are a special case of Lagrangian systems, for reasons in Section 4.7.)

Overall, the Newtonian statespace has a different structure from the Lagrangian statespace; not only that, it has more structure. This is another way of seeing the earlier conclusion that Newtonian mechanics possesses or requires more dynamical structure than Lagrangian mechanics; that the Newtonian dynamical laws presuppose more structure. We have arrived at this conclusion by means of two different routes. On one, we see that the Lagrangian equations are invariant under a wider range of coordinate transformations, which indicates that they require less structure. On the other, the statespace on which the Lagrangian equations are defined possesses less structure. This is an instance of a general phenomenon we saw in Chapter 2: there are two epistemic routes to learning about a given structure, either by means of invariant features under allowable coordinate transformations, or without reference to coordinates and their transformations.

The Newtonian statespace has more structure in a slightly different sense from some of the examples discussed earlier. A Euclidean plane with a preferred location or a preferred direction has more structure than a plane without, in that we can add such a structure to any Euclidean plane. The Lagrangian statespace has the structure of a Riemannian manifold, and not any such manifold admits a Euclidean metric (consider the surface of a sphere). What is more, since only a proper subset of Riemannian manifolds allow for a Euclidean structure to be added, it may seem as though the relationship between these formulations is not one of differing amounts of structure, but a narrowing down of the class of models.

how different conceptions of the theories lead to differing conclusions for their requisite structures. I claim that the above difference in structure holds on a natural and standard conception of the theories (which is not to say that this is the only possible conception). Barrett (2015a) also notes that one could simply stipulate that "physically reasonable systems" have regular Lagrangians, an idea that I am tempted by (for rheonomic systems, e.g., and more generally in the absence of the above assumptions, we cannot prove standard energy conservation (Lanczos, 1970, Secs. 1.8, 5.3)). At best, my arguments generalize; at worst, they hold for the standard theory and the "standard case" (Mac Lane, 1986, 286).

The formalisms do still presuppose different amounts of structure in various senses discussed earlier, however. Not any Lagrangian statespace can be given the structure of a Newtonian statespace. But the mere fact that not all Lagrangian statespaces admit a Euclidean metric does not stand in the way of their having less structure. The Lagrangian statespace has less structure in that we must first define a general type of (smooth manifold) structure that admits a Riemannian metric[51] before we can define a Euclidean metric, even though some spaces with that general structure won't permit a Euclidean structure to be added. The Newtonian statespace structure presupposes the Lagrangian type of structure, but not vice versa; the former is a special case or special kind of the latter structure. Compare the fact that a differentiable structure is intuitively more structure than a topology, even though not every topological space can be given a differentiable structure: only the topological manifolds can be endowed with that extra structure. Even so, a differentiable structure is more structure; a differentiable structure presupposes a topological structure, but not vice versa; a differentiable manifold is a special case or special kind of topological space. In fact, the comparison-of-models idea yields a similar conclusion. According to that idea, the models of Newtonian mechanics form a proper subset, they are a special case, of the models of Lagrangian mechanics, and in this sense one theory has more structure than the other, even though not every model of one can be given the extra structure of the other. I prefer not to put it this way, though, since I don't think the relationship between the theories corresponds to that of a subset of models, for reasons in Section 4.7.

(Recall from Section 2.4 that as we go up the hierarchy of structures, the associated group of mappings that preserve a given structure generally becomes narrower. This suggests that the size of the structure-preserving transformation (automorphism) group can be used to measure structure, with a larger group indicating less structure; that is, it seems as though we can use Klein's Erlangen program, which characterizes a given geometry via its transformation group, to compare different amounts of structure, by comparing the sizes of the relevant transformation groups. Often this is the case. However, it fails to work in the current case. For any Riemannian space of dimension n, the one with the largest group of isometries (for the given n) is a space of constant curvature; that is, a space with a Euclidean, spherical, or hyperbolic geometry. Comparing the sizes of the groups of structure-preserving transformations in this case does not give the right verdict on the spaces' relative amounts of structure. The reason for the failure is that Riemannian spaces in general lie beyond the scope of Klein's program. As Roberto Torretti says, "Klein's conception is too narrow to embrace all Riemannian geometries, which include spaces of variable curvature. Indeed, in

[51] Any smooth manifold (assuming the topology is Hausdorff and has a countable basis) can be endowed with a Riemannian metric (Burns and Gidea, 2005, 116).

the general case, the group of isometries of a Riemannian n-manifold is the trivial group consisting of the identity alone, whose structure conveys no information at all about the respective geometry" (2019). Klein's program simply "does not encompass Riemann's theory of metric manifolds" since for a general manifold of this kind, it is "powerless to characterize its structure" (Torretti, 1996, 283, n. 1).)

If you measure the structure of a space by the number of fundamental relations or predicates defined on it, as some philosophers seem to do, then the Lagrangian and Newtonian statespaces will appear to have the same amount of structure: both are metric spaces, just with different metrics defined on them. On my way of thinking, by contrast, the difference between the metrics itself indicates a difference in structure. The number of relations defined on a space is then not the final word on how much structure there is, for we must take into account the natures or definitions of the relations themselves. I do not have a detailed argument against the alternative, but will reiterate that my own way of measuring structure accords with an important mode of reasoning we often use in physics and mathematics, as we see by example here.

Jessica Wilson says that "energy-based" and "force-based" theories of mechanics are "mutually supporting, compatible perspectives on the phenomena of mechanical motions" (2007, 179), since their basic laws and principles are inter-derivable. Presumably, in her view, Newtonian and Lagrangian mechanics are equivalent in all relevant respects. In particular, they must either have the same structure, or else any difference in structure does not amount to a non-equivalence between them. Against this, I claim that there is a difference in structure required by their respective dynamical laws, which marks an important non-equivalence between the theories: a mathematical non-equivalence that reflects a particular physical non-equivalence, more on which in the next section.

Roger Jones (1991) agrees that these theories, interpreted realistically, yield different pictures of the physical world. However, he concludes that this is problematic for the realist, who will be forced to choose between different formulations that are (widely regarded as) mere notational variants, yet which differ physically if interpreted realistically. Since there seems to be no reason to choose one picture of the world over the other, the realist does not—cannot—know "about what to be a realist" (Jones, 1991, 190). As I see it, the realist should simply turn this argument on its head. There is no problem for the realist in this case, for we *should* interpret the theories differently. And once we do, we find principled grounds for choosing between them, which I turn to now.

4.7 Applying the minimize-structure rule

Since Lagrangian mechanics and Newtonian mechanics describe classical systems equally successfully, it seems as though the extra structure of the latter is superfluous.[52] The minimize-structure rule supports this conclusion. This rule says to use the minimal mathematical structure needed to formulate the (fundamental) laws, and to posit the correspondingly minimal physical structure in the world. Any structure beyond that is unnecessary, excess. Correlatively, given two different theoretical formulations, we should choose the one requiring less mathematical structure, and we should posit the lesser physical structure in the world.

We should therefore choose the Lagrangian theory and structure over the Newtonian theory and structure—both when it comes to the mathematical structure for the laws, and the physical structure of the world. By extension, these are not wholly equivalent theories. They are not mathematically equivalent, and they say different things about the nature of the physical world.

As with any principle governing scientific theory choice or interpretation, the minimize-structure rule holds ceteris paribus. If two theories do not do equally well predicting the phenomena, say, then we should not heedlessly choose the one with less structure. The competing formulations must do equally well in other respects before we try to minimize structure. A different way to put it: we should generally eliminate structure, but we had better be sure that it is genuinely excess. We mustn't eliminate any structure that is truly required by the physics.

In this case here, other things do seem to be equal. Lagrangian and Newtonian mechanics do equally well with respect to other theoretical criteria; textbooks tell us that they are (otherwise) equivalent. Our rule then says to choose the former. There could be reasons to infer the extra structure of Newtonian mechanics, just as there could be reasons to infer an absolute space structure for Newtonian physics or an absolute simultaneity structure for special relativity. In the absence of such reasons, we should not do so.

You might think the directness criterion I've mentioned, that we should prefer formulations that more directly represent the physical world, pulls in the other direction. Newtonian mechanics is formulated in terms of vector quantities, which directly mirror the nature of Newtonian forces. Lagrangian mechanics is formulated in terms of generalized coordinates, which needn't correspond to ordinary positions and velocities, and thus seem to only indirectly represent physical reality. The Newtonian formulation seems more direct and thereby preferable; or at the very least, it may seem as though we need to find a more direct formulation of Lagrangian mechanics before we can compare the theories' structures.[53] However, the role of generalized coordinates in the Lagrangian equations is really to serve

[52] Baez says that the Lagrangian equations get us "closer to the fundamental degrees of freedom of the system and so we cut out a lot of the wheat and chaff (so to speak) with the full redundant Newton equations" (Baez and Wise, 2005, 20).

[53] In fact, I see the directness criterion as relevant primarily to a choice among different formulations of a theory rather than a choice between different theories, whereas the minimize-structure rule applies to both—although there is a clear-cut way to compare structures across different theories, it is harder to see how to weight the relative directness of formulations that are said to be about different physical realities—but I won't argue the point here.

as placeholders for any set of coordinates we might use, thereby allowing us to encapsulate the dynamics in a way that abstracts away from any particular coordinates. In that sense the Lagrangian formulation is direct, characterizing the physics in a way that is independent of coordinates.

Not all mathematical differences between formalisms must correspond to genuine physical differences: there can be mere notational differences. Why not think that this is the case here? Because the mathematical formulation of the laws presupposes or requires a certain structure—this structure is required to support the laws, in the sense discussed in Chapter 3—and this structure differs between the two formulations. This suggests that something in physical reality correspondingly differs, just as we infer of candidate fundamental theories whose laws presuppose different spatial, temporal, or spatiotemporal structures. More generally, we should be realists about the structure required by the fundamental dynamical laws,[54] what I have been calling a theory's dynamical structure, in that this is something that we can be right or wrong about, depending on what the world is like, and which we should aim to minimize—a "structural realism" different from what currently goes by the name. (What would be a mere notational difference? Differences in mathematical formulation that don't amount to a difference in structure, as in the case of the Lagrangian laws stated in terms of different kinds of coordinates, or the Newtonian laws stated in terms of momentum as opposed to acceleration (note 5). Note however that there can be mathematical differences that don't amount to differences in structure, yet do amount to more than mere notational differences; reasons for this in Chapter 7.)

The structure required by Lagrangian mechanics is less than we ordinarily think. We usually think of the classical mechanical laws as presupposing a Euclidean metric, and the physical space of any classical world as being Euclidean in nature. This is because we usually think of classical mechanics as equated with Newtonian mechanics, which does presuppose this structure. But the mathematical formulation of the Lagrangian laws does not presuppose a Euclidean metric. Hence the theory's physical space, whose structure is coded up in the mathematical structure needed to support the laws (again compare our usual spacetime inferences) does not have a Euclidean structure. Lagrangian mechanics endows physical space with a more general, locally defined distance measure, not a globally defined Euclidean one. There is (affine) structure to distinguish between particle trajectories that are straight and those that are not, but the former needn't be straight lines in the sense of Euclidean geometry.[55] Putting this in the four-dimensional spacetime

[54] Again, as with the principles from Chapter 3, I leave it open whether something like this holds for nonfundamental laws.

[55] Maudlin says that, "Explication of Newton's Laws of Motion ... does not make any essential use of the Euclidean structure of the space: although Newton presumes space to be E^3, nothing in his dynamics requires this" (2012, 34). I would say this instead: although Newton assumed this spatial structure, and the laws as he saw them require it, much of their physical content does not really need it, as evidenced by

terms prevalent in current philosophical discussions: the spacetime of Lagrangian mechanics is a neo-Newtonian spacetime in which the spatial surfaces of constant time are generally curved Riemannian manifolds rather than flat Euclidean spaces. (In Chapter 7, we will see reasons to question whether this is a fully equivalent way of putting the conclusion.)

Now we can see why I said that the relationship between these theories does not correspond to that of a subset of models. The minimize-structure rule tells us that according to Lagrangian mechanics, physical space does not have a Euclidean metric structure. Models or worlds in which space has this structure are then not models or worlds of Lagrangian mechanics. In other words: Newtonian mechanics and Lagrangian mechanics are distinct theories, with distinct sets of models, which say different things about the physical world (while having enough in common to both qualify as "theories of classical mechanics"). Compare: in special relativity one might posit an absolute simultaneity structure, for a Lorentzian version of the theory, or one might not do this, for an Einsteinian version. It is reasonable to consider these distinct theories, with distinct sets of models, which say different things about the fundamental nature of the world (while having enough in common to both qualify as "theories of special relativity"[56]). Notice that we can compare structures even so: Lorentz's theory has more structure (in both the formalism and in the world) than Einstein's theory, even though the models of one theory do not form a proper subset of the models of the other.[57]

You may balk at drawing conclusions about the nature of physical space from a theory's statespace, as I do along one route to the above conclusions. A statespace seems purely abstract, "having nothing to do with the physical reality" of a system, being "merely *correlated*" with it (Lanczos, 1970, 13; original italics). However, notice that I am not claiming that the configuration space (nor the statespace as a whole) *is* a world's physical space, the "container" for material objects. Nor do I claim that a theory's statespace is directly isomorphic in all respects to physical space. Configuration space typically has more than three dimensions, for instance, whereas the physical space of a classical mechanical world (by initial assumption) is three-dimensional. The claim is rather that the statespace structure encodes, and thereby tells us about, the structure of physical space. (This is the sense in

the success of Lagrangian mechanics, which does away with it. (Elsewhere Maudlin notes that Galilean spacetime, which assumes a Euclidean spatial structure, is "the ideal arena for Newtonian mechanics" (2012, 65).)

[56] As mentioned in Section 3.3, you might not want to consider Lorentz's a theory of special relativity, since it posits structure that is not Lorentz invariant, contrary to the spirit of the principles of special relativity. That depends on which principles one takes to be the core principles of special relativity. Regardless, the theory differs from Einstein's in the above way.

[57] An alternative is to say that Lagrangian and Newtonian mechanics are equivalent (only) in Euclidean space. I do not want to go that route. That would be like saying that Einstein's and Lorentz's theories are equivalent (only) in a spacetime with an absolute simultaneity structure. Better to say that Einstein's theory simply does not posit an absolute simultaneity structure.

which we should be realists about a theory's statespace structure, not in that we should take the statespace to *be* physical space.) Just as any system's configuration space represents the physical space its particles move around in, so too a world's configuration space represents the physical space its particles move around in.

You might reply: but given that we can use coordinates that don't resemble ordinary position coordinates, why think that configuration space in any way represents the structure of physical space? Answer: because that is what configuration space *is*, a mathematical space in which we represent the possible positions of particles in ordinary physical space, their possible positions or spatial configurations: "it is [or represents] the set of all possible positions" (Singer, 2001, 19). That is the very idea or essence or definition of a configuration space.

A related worry. The points in a statespace represent the different possible states of a system. Even if a metric on this space satisfies the formal mathematical definition of a distance function, it does not seem to be a genuine *distance* measure in the sense we usually mean for a true physical space or a geometrical space we might use to represent it. The statespace metric is rather a way of formalizing the relative similarity of different physical states: points that are "close" to one another in the statespace represent states that are physically similar to one another. The statespace is merely a "metaphorical space," in the phrase of Maudlin (2014a). So why think the metric on this space tells us anything about a distance measure on a genuine physical space? You might say more generally that the formulation of a theory in terms of its statespace is "mathematically useful but not metaphysically germane," as Wallace puts it, since the "metaphysics of a theory, presumably, is to be understood in terms of the actual properties and relations holding between the objects that make up the world according to the theory, and state space is just an abstract mathematical tool" (2013, 205). A statespace in general, its metric structure in particular, are mathematical tools with no direct bearing on physical reality.

We can grant the thought that fuels the concern—that the statespace is not itself a part of physical reality, and that not all its features directly represent that reality—while still allowing that it is used to represent, and can therefore tell us about, physical reality.[58] As Wallace notes, a classical statespace is constructed on the basis of assumptions about the (meta)physics, such as that the theory is about n particles in three-dimensional space. The structure of the statespace is not completely irrelevant to a theory's metaphysics (even if it represents it somewhat indirectly; more in Chapter 7). Nor is the similarity metric wholly unrelated to the distance measure on physical space. States whose particles are near each other in physical space will be represented by points that are near each other in the

[58] Ismael and van Fraassen note that the reason we care about a statespace—or a "space of possibilities"—in physics is that it is "relate[d] in a principled way to the structure of actuality" (2003, 385).

statespace, and in this way the similarity metric is coding up facts about physical distances.

You might say that the statespace gets its structure from physical space, which we know antecedently to be flat and Euclidean: that the $3n$-dimensional base space (on either theory) "carries a flat metric which it inherits from the metric on physical space" (Belot, 2000, 572). Yet although the statespace does inherit its structure from physical space, we don't know the structure of physical space a priori. Nor do we directly observe it. Rather, we infer this structure from the theory that best explains the motions and interactions of physical objects. (Compare Albert (1996, 2015, 2019a) on inferring the structure of physical space from the dynamics.) The fact that the structure of configuration space depends on the structure of physical space doesn't interfere with the fact that our evidence for the latter can come from the former, and from the dynamics more generally.

Although I claim that the metric on configuration space represents the metric of physical space reasonably directly, I do not deny that physical space, on either theory, is three-dimensional, whereas the configuration space has $3n$ dimensions (for n particles). The dimensionality of physical space is more indirectly represented in the statespace structure (for more than a single particle) than the metric is. Whence the difference?

The difference is that it is an initial physical postulate of the two theories, as I am discussing them here, that there are n particles in three-dimensional space. It is stipulated from the outset that each theory is about n particles moving around in three-dimensional physical space, and the statespace is constructed on the basis of this stipulation. What is more, nothing about the dynamics leads us to consider revising this assumption. (Albert (1996, 2015, 2019a) argues that the dynamics of quantum mechanics does indicate a fundamental physical space of a very different dimensionality.) The metric structure of physical space is not similarly posited or known from the outset: for that we must turn to the dynamics.

This helps address a question you might have about how the current discussion interacts with the matching principle from Chapter 3. Doesn't the matching principle tell us to posit physical structure in the world that matches the mathematical structure needed to support the laws; so that if a particular statespace structure is needed for the laws, we must conclude that the physical structure of the world directly matches everything about the statespace structure? The answer is no, and the reason is the initial physical posits being assumed; namely, that both theories under consideration are about particles moving around in three-dimensional physical space. Relative to this assumption, certain mathematical features of the statespace (like its dimensionality) are not candidates for a direct match with physical reality, but are encoded in the statespace structure more indirectly. This still leaves it open for other features (like the metric) to more directly match the structure of physical reality.

The examples discussed in Sections 4.3 and 4.5 were (idealized to be) of single-particle systems. In Newtonian mechanics, multiple particles interact by means of forces exerted across Euclidean spatial separations. This may make you wonder whether Lagrangian mechanics, assuming that it does away with a Euclidean metric, can handle systems of many interacting particles. It can, using its well-defined notion of geodesics given by the Riemannian metric. A Euclidean metric is not needed. Hence (again) we should infer that this is not part of the structure of a Lagrangian world. Relatedly, we should infer that Newtonian forces are not fundamental to such a world. Where in Newtonian mechanics, force "plays the role of a fundamental primitive in dynamics" (Sklar, 2013, 111),[59] in Lagrangian mechanics, that role is played by the Lagrangian energy function. (I leave it open whether the Lagrangian is absolutely fundamental, or whether instead the potential and energy functions are; or the (generalized) particle positions and velocities out of which these functions are defined; or whether these might all be equally fundamental. What matters here is the contrast with the Newtonian emphasis on forces.)

It took the case of the pendulum, idealized as a constrained single-particle system, to bring out the core differences between the theories. However, you might think that only the behavior of free—unconstrained—particles can be a guide to the true nature of a world's physical space. (Perhaps because constraints themselves do not seem to be fundamental things.) And you might worry that if we do limit our focus to the behavior of free particles, then the two theories will indicate the same kind of physical space after all.

I take it that in general, the behavior of free particles is not enough to convey a complete picture of the world according to a physical theory. We also need information about particle interactions, which in the case of the pendulum are being idealized as constraints that operate external to the system. (It is noteworthy that in Albert's discussions (1996, 2015, 2019a), the interaction terms of the Hamiltonian are what indicate the nature of space.) Only a theory's full dynamical structure, including that underlying interactions, can tell us about the complete physical structure of the world.[60] In Newtonian mechanics, for example, it was crucial to pay attention to the nature of forces, which only arise from interactions between particles. (On this theory, forces, and the lack thereof, are needed to indicate which particles are on straight trajectories, which motions are inertial or force-free, after all.) There are certainly questions that remain about how exactly to understand the nature of constraints and the other idealizations we standardly

[59] Sklar is talking here about the entire Galileo–Huyghens–Newton tradition in mechanics.

[60] You might think that in sparse worlds containing only a single particle, or worlds that can be said to contain only non-interacting particles, we may infer things about the spatial structure without knowing anything about interactions, but those are special cases; and even there, we arguably need to know something about how a particle would behave if it interacted with other particles in order to infer what the spatial structure is really like.

make for systems like the pendulum, but these are things I aim to leave aside here: regardless, we can see the difference in how each theory treats such a system, which forms a basis for my conclusions. For that matter, although the theories' differing treatment of the pendulum helps crystallize the point, the favoritism of the Newtonian theory for rectangular coordinates can be seen in the case of free motion, too, simply by inspection of equations 5 and 6 (inertial motion will not be described by $\ddot{r} = 0$ and $\ddot{\theta} = 0$; the equation will contain further terms, including first derivatives); so that even the case of free motion should support the above conclusions. By studying a system like the pendulum, we mange to elicit the underlying physical reasons for the difference.

In Lagrangian mechanics, the fundamental dynamical quantity that specifies a system or world is the Lagrangian, which contains information about the number and types of particles and determines the equations of motion: "A mechanical system is completely described by its Lagrangian" is how one book puts it (Hamill, 2014, 27).[61] (That is, given the initial postulates mentioned above, the Lagrangian fully describes a system; the caveat since on its own, the Lagrangian won't determine the number of particles and the dimensionality of the space they are in, for reasons discussed by Albert (2019a).) The theory does not fundamentally distinguish systems that are alike in these things.

This means that, according to Lagrangian mechanics, there is a fundamental difference between a single particle in free fall and a single particle (with the same intrinsic features) in simple harmonic motion: different Lagrangian and equation of motion (recall equations 8 and 9 and note 28). But consider a single particle undergoing simple harmonic motion along a straight line, with equation of motion $\ddot{x} = -kx$ (x the displacement along the line), and a single particle with the same intrinsic features undergoing simple harmonic motion along an arclength, with equation of motion $\ddot{\theta} = -k\theta$ (θ the displacement along the path).[62] Lagrangian mechanics views these two systems as on a par: there is no fundamental difference between them; they are simply being described in different ways. The only difference is whether the Lagrangian function and equation of motion contain rectilinear coordinate x or angular coordinate θ, which is not a genuine difference on this theory. The theory does not recognize such differences, not without adding further structure beyond what the laws require, which our guiding principles tell us not to do. (The form of an equation is not the final word on what it represents. A given (form of) equation can describe different physical systems depending on how we interpret it. You might conclude that the two systems are different on the grounds that we interpret the equations differently. Although we could make certain assumptions that yield this result (more on this

[61] There can however be different but equivalent Lagrangians for a given system (José and Saletan, 1998, Sec. 2.2.2).

[62] I am grateful to Adam Elga for prompting me to think about this example.

idea in Chapter 7), I have not done so here: instead, we have learned that there is no real difference between them.)

By contrast, Newtonian mechanics does distinguish these systems, for the laws distinguish between different kinds of coordinates. The laws privilege Cartesian coordinates, a preference that presupposes a Euclidean spatial structure, which distinguishes straight Euclidean lines from arclengths. Another way to see it is that the forces in each case will be different, and only Newtonian mechanics considers this to be a fundamental difference. Once again, Lagrangian mechanics contains less structure, for it fails to draw certain distinctions, to recognize certain notions, that Newtonian mechanics does.

Take what we might call the "absolute structure" of a theory: the structure the different worlds governed by the theory all have in common.[63] In Lagrangian mechanics, this includes a metric structure, but not a Euclidean metric structure in particular. This theory can say, of any world it describes, whether a particle is traveling on a locally straight line (geodesic) or not, since according to the theory the world will have the requisite structure to support this distinction: this is part of the theory's absolute structure. But the theory cannot say, of a given world, whether a particle is traveling on a straight Euclidean line or not, for according to the theory there isn't the structure necessary to support this distinction: this is not part of the theory's absolute structure. This allows us to say that the theory does not fundamentally distinguish between the two systems above, on the grounds that the theory's absolute structure does not support the distinction. On the other hand, a Euclidean spatial structure is part of the absolute structure of Newtonian mechanics, possessed by any world governed by the theory. This theory can say, of any world it describes, whether or not a particle is traveling on a straight Euclidean line. Arguing for all this further, however, requires more of an account of fundamentality and absolute structure than I have to offer here.

Reichenbach (1958) says that there is no fact of the matter about whether the space of a given world is flat and there are undetectable universal forces in certain regions, or whether space has non-zero curvature in those regions (assuming these give rise to the same observational evidence), independent of a conventional decision about whether to allow for universal forces. You might for similar reasons think that there is no real difference between the structure of a Newtonian and a Lagrangian world. Against this, I assume that a world's spatial structure is not up for conventional grabs: there is a fact about this (more in Chapter 5). Alternatively, you might think that different physical descriptions that seem to disagree on whether a world's space is flat can in fact "say the same things about the world" (Weatherall, 2016a, 1087). Against this, I assume that such descriptions do say different things about the world (subtleties in Chapter 7). If differences in

[63] Compare and contrast the absolute objects of Anderson (1967); Friedman (1983).

spatiotemporal structure do not count as genuine differences between worlds, then I am not sure what do.

In this case here, considerations of comparative structure choose the right theoretical description. Lagrangian mechanics has all the structure that classical mechanics needs. Hence (ceteris paribus[64]), this is all the structure we should infer a fundamentally classical mechanical world to have. In other words, Lagrangian and Newtonian mechanics are not wholly equivalent theories. A surprising conclusion, warranted by a familiar general rule.

At this point you may be left wondering whether there can be any cases of theoretical equivalence in physics. I return to this in Chapter 7.

[64] This is not to rule out the possibility of a formulation that requires even less structure: North (2009).

5
Spatiotemporal Structure

> For these computations, progressing by means of arithmetical operations alone, very often express in an intolerably roundabout way quantities which in geometry are designated by the drawing of a single line.
>
> Isaac Newton

5.1 The debate about spacetime

Consider the following question: does space (in traditional terms) or spacetime (in contemporary terms) exist, over and above material bodies? The substantivalist says that it does. The relationalist says that it doesn't. The relationalist believes there are material bodies related to one another spatiotemporally, but there is no further thing in which those relations inhere, no "container" in which material bodies are located.

This is a longstanding debate that many contemporary philosophers believe has stagnated, become non-substantive, or is purely metaphysical, entirely divorced from the concerns of physics. For instance, consider a conception of the debate that harkens back to Leibniz and Newton, as the question whether space is a substance. Opponents on either side of this question may seem to differ in what they mean by a substance, so that "the putative opponents fail to express contrary positions" (Myrvold, 2019, 139) and wind up simply talking past each other.

Many philosophers have concluded, for a variety of reasons, that the traditional dispute is irrelevant to physics, either in general or in the context of particular theories or situations (examples in the footnote).[1] Robert Rynasiewicz (1996) argues that even if the debate was once relevant to physics, it is no longer: it has become a merely verbal dispute over which things to call "space" versus "matter", with no objectively correct answer to be had. David Malament wonders whether there is any clear-cut dispute between the relationalist and the substantivalist, given how unclearly their views are typically formulated: "Both positions as they

[1] Claims to that effect can be found in Stein (1970a, 1977b); Malament (1976); Horwich (1978); Friedman (1983, 221–3); Earman (1989); DiSalle (1994); Leeds (1995); Rynasiewicz (1996, 2000); Belot and Earman (2001, Sec. 10.7); Dorato (2000, 2008); Pooley (2013, Secs. 6.1.2, 7); Curiel (2018); Slowik (2016); Myrvold (2019). Earman (1989) advocates the need for a *tertium quid*.

Physics, Structure, and Reality. Jill North, Oxford University Press (2021). © Jill North.
DOI: 10.1093/oso/9780192894106.003.0005

are usually characterized . . . are terribly obscure. After they are qualified so as to seem intelligible and not too implausible, it is hard to retain a firm grasp on what divides them" (1976, 317). For that matter, there are so many different conceptions of the dispute available in the literature that you might conclude that there simply is no overarching, well-posed question in the vicinity (Curiel, 2018). Belot sums up the situation by remarking on "the fragile health of the substantival-relational debate" (1999, 38).

In this chapter, I discuss this debate. I explain why one might reasonably feel that this debate has stagnated, while suggesting that there is nonetheless a dispute in the vicinity that is both substantive and relevant to physics. Although the conception of the debate that I will propose is not exactly the traditional one, it is very close to it in spirit, capturing the same core ideas. Indeed, it is essentially that dispute, updated to take into account more recent developments in physics and in philosophy. At the same time, this understanding of the dispute allows us to formulate the most plausible (if not entirely traditional) versions of the two main positions on the issue, with a clear point of disagreement between them. Finally, we will see that putting things in this way unearths a new kind of argument for substantivalism, or at least a new challenge to relationalism, given much of current physics, while at the same time allowing for future physics to indicate otherwise. In that way, this dispute will continue to remain relevant to future developments in physics.

You may wonder what all this has to do with structure. One of the reasons for reframing things as I suggest is in order to update the traditional dispute to the terms of modern physics, which means putting things in terms of spatiotemporal structure. And once we put things in these terms, the various considerations about structure, and the principles governing our theorizing about it, will kick in. In all, a seemingly subtle shift yields surprising progress on a longstanding issue that many people feel has stagnated.[2]

5.2 A brief history of space

I begin by considering some examples from the traditional dispute (that is, from the beginnings of the modern debate, usually traced to Newton and Leibniz and Clarke) that continue to be discussed. I do this for a couple of reasons. First, these examples can leave one with the feeling that there isn't much at stake in the dispute, since each side seems to have adequate responses to the cases aimed against it. This bolsters the feeling that the traditional debate is at an impasse, as well as the thought that there may be no real difference between opposing sides, and helps motivate the project of locating a new understanding of the dispute. Second, the

[2] Related discussion is in North (2018). This chapter includes an overview of the traditional debate and streamlines aspects of the discussion, leaving further details to the earlier paper.

replies that I suggest on behalf of each side (replies that can for the most part be found in the literature, though not always in the ways I will put them) are not exactly the traditional ones. Rather, they hint at a new way of understanding the debate and the main positions on it. My aim in this section, then, is not to rehash historical details (nor to be entirely historically accurate), but to move us toward a better understanding of the debate, one that is motivated by thinking through some of the classic cases. So even if the dispute starts to feel inconclusive, and somewhat elusive, in the face of traditional arguments, we will see that they ultimately point to a new understanding of the dispute, which is both substantive and relevant to physics.

In keeping with the historical debate, I will discuss these examples primarily in terms of the question of the existence of space, although I will start to put some things in spacetime terms. (In the next section, I discuss things entirely in those terms.) I will also assume a Newtonian physics, as was assumed at the time these examples were proposed. As we proceed, notice how the examples suggest that the question of the existence of space is intimately bound up with the aim of explaining objects' motions. The debate about the ontology is a debate about what we need to assume or posit in the physical world in order to account for objects' motions.

I am going to spell out the two main views on the issue in a particular way later in this chapter, but for now think of them like this. According to the relationalist, all that exists in the physical world are material bodies that enter into various spatial relations. (Temporal relations too, but set this aside for now.) The spatial facts about a world are about these objects and the spatial relations among them. According to the substantivalist, there is a physical space in addition to material bodies, and the spatial facts about a world are about this space. (In the terms of Newton and Leibniz, the question at issue is whether space is a substance, some kind of independently existing thing.)

Consider that famous example: Newton's bucket experiment. Newton used this to argue for substantivalism about space. He argued that empirical evidence reveals that absolute motion, hence absolute space, exists—where absolute motion, in his terms, is motion relative to an absolute persisting space rather than other material bodies. He argued that there are observable effects of absolute motion, which demonstrate the existence of space.[3]

Newton's experiment is this. A bucket of water is suspended from a rope. Twist the rope and let the bucket spin as the rope unwinds. What happens? Initially the bucket spins while the water remains still—there is relative motion between the

[3] I won't discuss the question of how best to construe Newton's own arguments. Nor will I enter into disputes over whether Newton really was a substantivalist (and Leibniz really a relationalist). Newton did not think that space was a genuine substance, in his sense of the term, but he can still be seen as a substantivalist in a relevant sense. (In its contemporary incarnation, "substantivalism" is not meant to correspond to any particular historical notion.) See Stein (1970a, 1977b); DiSalle (1994, 2002) for a different take on the historical record.

water and the bucket—and the surface of the water is flat. Soon afterward, the water starts to rotate with the bucket. The water's surface develops a concave shape, at which point there is no relative motion between the water and the bucket. After that, the bucket will gradually slow down to a stop, while the water keeps spinning with a concave shape: there is relative motion between the water and the bucket. Eventually the bucket and water both stop, and the water's surface is flat. There is once again no relative motion between the water and the bucket.

Question: why does the surface of the water become concave? The answer seems clear. The water's surface is concave because, and when, the water is spinning. But spinning with respect to what? It is in trying to answer this question that the relationalist seems to be in trouble. The problem is that the surface shape of the water does not vary in the right way with the relative motion between the water and the bucket. Sometimes the water's surface is flat while the water is spinning relative to the bucket, sometimes it is flat while the water is not spinning relative to the bucket. The relationalist cannot say that the water is spinning relative to a background space, since there is no such thing. The water must be spinning with respect to some other material body, yet the bucket clearly won't do the job.

Newton concludes that the relationalist cannot explain the phenomena, observable phenomena that even the relationalist will acknowledge and want to be able to explain. The water must instead be spinning relative to space itself. More generally, the existence of an absolute space—an unchanging, persisting space that exists over and above material bodies—seems to be needed to account for the observable effects of non-inertial, or accelerated (in this case rotated) motion, such as the shape of the water in the bucket.

Now, Newton's argument really only shows that the explanation of the phenomena cannot come from the motion of the water with respect to the bucket. (It helps to keep in mind that Newton was arguing against Descartes' brand of relationalism in particular.) The argument does not show that there is no material object with respect to which the water could be spinning. Ernst Mach, for one, said that the water is spinning relative to the fixed background stars.[4] The stars essentially form a global inertial frame relative to which objects can be rotating or not, with no need for an absolute space to account for facts about rotation. There are oddities of a physics that results from such a view, which I won't discuss here. (Mach himself did not offer such a theory, but more recent work has shown ways in which it could be developed.)[5] The main point for us is that, those oddities aside, there are plausible enough replies to Newton's challenge. The relationalist simply needs

[4] Mach (1960); excerpts in Huggett (1999, Ch. 9).

[5] Even without a fully developed theory, we can note some odd features it would have, such as being highly non-local and requiring the total angular momentum of the universe to be zero. (Newton dismissed such a theory in his *De Gravitatione*, excerpts in Huggett (1999, Ch. 7).) Barbour and Bertotti (1982); Barbour (1982, 1999) contain one recent Machian theory.

to find some other material object(s) relative to which the water can be rotating or not.

You might think, with Earman (1989, Sec. 4.1), that this is not enough. To reply that it is possible to explain certain phenomena by reference to material bodies is insufficient. The relationalist needs to show that the resulting physical theory is at least as good in predictive and explanatory power as the substantivalist's. And it seems that what Newton has shown is precisely that the relationalist's physics is unlikely to be as good, for the substantivalist has a particularly simple and natural explanation of the phenomena. Since the various relationalist theories that have been proposed remain contentious at best (note 5), perhaps it is a mistake to say that the relationalist has a "plausible enough" reply at this stage. Although I am more optimistic that a relationalist mechanics can meet Earman's challenge, I agree that there is some cause for concern. At the end of this chapter, I will argue that a different, not unrelated, worry plagues the relationalist's physics. For now, though, let's grant the relationalist the benefit of the doubt. Let's grant that, at the least, the relationalist has a reply to the case of the bucket, and if not yet a fully worked-out theory, at least the beginnings of one.

A variant of the case creates more trouble for the relationalist, however, for it removes all possibility of explaining things by reference to other material bodies. (This too comes from Newton.) Think of two rigid globes attached by a rigid cord in an otherwise empty universe. Two different types of motion seem to be available to this system: it could be rotating about its center of mass or not so rotating. Put another way, there seem to be two different possible worlds: in one, the only material thing that exists is this system and it is rotating; in the other, only an otherwise-identical system that is not rotating. Newton says there will be an empirically detectable difference between the two: in the rotating system, but not in the non-rotating one, there will be an observable, measurable tension in the connecting cord. (Different tensions moreover correspond to different directions and rates of rotation.)

It seems as though the relationalist cannot explain the difference between the two situations or worlds. The relationalist facts are the same in each: there is no relative motion between any material bodies in either one. According to Newton, by contrast, the difference between the two stems from their different states of motion: there is a tension in the connecting cord between the rotating globes, which is absent from the non-rotating globes. Newton concludes that a background space is needed to account for this difference: in one situation, but not the other, the globes are rotating with respect to space.[6]

[6] You might think the example beside the point. Mach said that it is "not permitted to us to say how things would be" if the world were other than it is (quoted in Earman (1989, 83)). Against Mach, typical physical theories have lots of things to say about non-actual situations—the laws say what would happen under various initial conditions that do not actually obtain—and it is unclear that a viable physics could do away with all such claims. In any case, I will ultimately suggest a way of construing the debate that need not center on such examples.

Again there are various relationalist responses available. I want to mention one type of reply, which employs a strategy Oliver Pooley (2013, Sec. 6.1) calls "enriched relationalism." The relationalist can account for the distinction between the spinning and non-spinning globes by countenancing *enough* relations between material bodies. Maudlin (1993, 187) notes that the relationalist can account for the difference between the two situations by positing cross-temporal spatial distances between different temporal stages or parts of the system. These relations will be different for the two types of motion. (Maudlin calls this "Newtonian relationalism," since it effectively posits all the spatiotemporal relations that inhere in a Newtonian spacetime structure.) But the relationalist does not have to go so far as to countenance such distance relations. All that's needed are (affine) relations between different temporal stages or parts of the system that indicate whether the globes' worldlines are straight or curved.[7] (We might call this "Galilean relationalism," since it effectively posits all the spatiotemporal relations that inhere in a Galilean spacetime structure. I say this doesn't go as far as Newtonian relationalism for the sorts of structure-comparison reasons from Section 2.4.) These spatiotemporal relations differ in each situation, corresponding to the worldlines being curved in the rotating world and straight in the non-rotating world. The two worlds then do differ in relationalist facts.[8]

The above is starting to move away from a wholly traditional relationalism. The traditional debate, in keeping with the physics of the time, was about the existence of space and time separately. In that context, any spatial relations are assumed to hold between material bodies at a time, and more generally there is no recognition of essentially spatiotemporal relations. What we are seeing is that once we move to a spatiotemporal viewpoint, and countenance fundamentally spatiotemporal relations as well (such as the cross-temporal spatial distance relations of the Newtonian relationalist or the straightness-of-spatiotemporal-trajectory relations of the Galilean relationalist), the relationalist should have an adequate reply to the case of the globes as well as the bucket. Such a relationalist should generally be able to say whether a system is rotating or not, even in a world devoid of other material bodies. All of this is still in the spirit of traditional relationalism, however, since the phenomena are explained solely by reference to material bodies and the relations among them (including relations between different temporal stages of a given material body or system), without any reference to a background space.[9]

[7] Maudlin (1993, Sec. 4) notes such a view but immediately cites problems for its predictive power. For now consider this an initially promising response, if not yet a fully worked out theory. (Modal relationalism, Section 5.5 below, may evade the worry.)

[8] Some other replies: the relationalist can simply deny that it is possible for the globes to be spinning in the absence of any other material bodies; or deny that there is a fact of the matter about whether they are rotating or not in an otherwise empty universe; or claim that there is a brute difference between the two situations, say via a primitive monadic property of acceleration, not defined relative to other material bodies (Sklar, 1974, 230–1).

[9] You might think, and traditionally it was thought, that relationalism about space must go hand in hand with a relationalism about motion. We are starting to see that the question about motion comes

I haven't said exactly how the relationalist can countenance the requisite spatiotemporal relations: I turn to that below. The point for now is simply that phenomena involving rotation on their own do not immediately doom the view. They simply spell trouble for certain versions of it, namely any "impoverished relationalism" that fails to posit enough relations between material bodies. (This does tell against one traditional version of relationalism in particular, often called "Leibnizian relationalism," which eschews the above kinds of cross-temporal relations. I am suggesting that a viable relationalism must countenance some such spatiotemporal relations, to account for phenomena involving rotation, as well as for more general reasons to come.)[10]

Next consider the kinematic shift argument, aimed against the substantivalist. (This argument traces back to Leibniz.[11]) Consider our world and another just like it but with every material body moving with a different constant velocity throughout history. It is a consequence of Newton's laws, which are invariant under uniform velocity boosts, that these worlds are in principle empirically indistinguishable. The substantivalist seems committed to an unpalatable array of empirically indistinguishable yet distinct physical situations, since presumably there are facts about how fast objects are moving with respect to space, facts that render the original and shifted worlds physically distinct. The relationalist instead sees the shifted situation as merely a redescription of the original one, since the relationalist facts are the same in each case. This seems like the right result.

There is a way for the substantivalist to agree with the relationalist's conclusion, however. What the example reveals is that absolute velocity does not matter to the physics: this quantity is not evident in the phenomena, nor is it recognized by the physics governing the phenomena. This is what underlies the claim that the shifted situation will indeed "look the same." The relationalist can easily accept this. But the substantivalist should be able to as well, by adopting a conception of space on which the uniformly shifted situation counts as a mere redescription of the original one, just as the relationalist says. This would mean abandoning Newton's idea of absolute space and concomitant notion of absolute velocity, but it needn't mean abandoning substantivalism altogether. There may be a substantivalist view which says that space exists, while denying that there are facts about how fast objects are moving through space.

apart from the question about the ontology. This is one way that my understanding of the debate departs from tradition. Other contemporary discussions, such as Friedman (1983); Earman (1989); Belot (2000), also distinguish these questions.

[10] There are versions of enriched relationalism available for special relativity (Earman (1989, 128–30); Maudlin (1993, Sec. 5)) and general relativity (which may require a modal relationalism: Section 5.5 below). Earman (1989); Maudlin (1993, Secs. 6–7); Pooley (2013, Sec. 6.1.2) are more pessimistic about the prospects for an enriched relationalism in general relativity.

[11] And Clarke, who used this to argue *for* substantivalism. The "kinematic shift" and "static shift" labels are from Maudlin (1993). I won't discuss Leibniz's own arguments, which rely on particular metaphysical principles that are not needed for the main idea.

Recall from Chapter 3 that Newton can be understood anachronistically as postulating a Newtonian spacetime, which has the structure to support his notion of absolute space, a space that persists through time, relative to which there are absolute facts about objects' velocities and locations. Call such a view "Newtonian substantivalism." The suggestion is that the substantivalist should instead adopt a "Galilean substantivalism," believing in the existence of a space(time) that does not have the structure to support facts about absolute velocity—that is, a Galilean spacetime. Such a substantivalist avoids the charge of recognizing more physical possibilities than there are: the kinematic shift in particular does not yield a genuinely distinct situation. A Galilean substantivalist does not believe in the particular kind of space that Newton did, but does believe in something appropriately similar to it.

I have not said exactly how this kind of substantivalist assumes something appropriately similar to Newton's substantival space; I return to that below. The point for now is simply that the kinematic shift on its own does not immediately doom the view. Instead, the moral is that the substantivalist should be careful to posit the right structure to space(time), in this case a structure that recognizes the same possibilities the laws do.

This conclusion is interestingly analogous to the earlier one we saw for the relationalist. Just as the relationalist, in the face of the bucket and globes, should be careful to posit the right spatiotemporal relations needed to account for the phenomena—where the bucket and globes each suggest that there must be enough relations to undergird facts about rotation—so too the substantivalist should be careful to posit to spacetime itself the structure that is needed to account for the phenomena, where the kinematic shift suggests that absolute space and absolute velocity are not needed. (It is worth noting that the kinematic shift seems problematic for the substantivalist only given dynamical laws that are invariant under uniform velocity boosts; similarly, only given phenomena that are empirically indistinguishable under such a transformation. If the laws and phenomena were not so invariant, then we should conclude that the shifted world is distinct: it would be distinguishable from the original one. More generally, which facts are recognized by the physics, and so which situations or worlds count as distinct (because of differences in those facts), depends on the laws and the quantities in terms of which they are formulated.)

Another example in the traditional debate is the static shift, also aimed at the substantivalist. Think of our world and another one just like it but with every material body shifted over a uniform spatial amount throughout history. The relationalist thinks that these are the same world, differently described. The substantivalist seems committed to saying that these are distinct worlds or possibilities, since things are located at different places in space. Yet it seems as though there can't be any genuine difference between situations related by a global spatial shift.

Here is one type of reply.[12] The substantivalist, as much as the relationalist, can deny that the shift generates a distinct possibility. As Hartry Field (1985, n. 15) notes, one might reasonably think that the identification of objects across different worlds, including such objects as spacetime points or regions, is tied to their qualitative features, so that the original and shifted cases do not count as genuinely distinct. Just because you believe in the existence of space does not mean that you must be committed to trans-world facts about the identification of points or regions independent of their qualitative features, such as which objects are located at them. Using this idea, even the substantivalist can say that the original and shifted worlds are really the same world, differently described: shift the objects around uniformly, and you shift the points or regions with them. To put it another way, substantivalism on its own does not commit one to the view that non-qualitative differences between worlds are real differences. The substantivalist can adopt an anti-haecceitism according to which qualitatively identical worlds, such as worlds that are said to differ only in which objects are located at which points, are identical, full stop.

The above goes against one familiar take on substantivalism, according to which the view is committed to the static shift's generating a distinct possibility. John Earman and John Norton (1987) call this the "acid test" for substantivalism. But it nonetheless seems like a reasonable reply. It is in keeping with other kinds of conclusions we draw in physics. Given the symmetry of the physics—given that the laws and phenomena are invariant under uniform spatial translations—it is reasonable to conclude that the static shift does not generate a physically distinct situation: the original and shifted worlds are the same in all the ways that matter to the physics. And given the idea in the last paragraph, it seems as though the substantivalist should be able to say this as much as the relationalist can. This is a reasonable reply that, contra Earman and Norton, needn't mean rejecting substantivalism wholesale. It simply means adopting a view of modality according to which the shifted description represents the same possibility. (As Carolyn Brighouse (1994) argues, similar reasoning can be used to reply to Earman and Norton's (1987) hole argument. I won't be discussing the hole argument, since it does not introduce considerations relevant to us beyond those already revealed by the classic shift arguments.[13])

[12] Another: say that the possibilities are distinct but distinguishable, by ostension (Teller, 1991; Maudlin, 1993): we know that we are *here* (in this world, at this location) and not *there* (in that world, at that location) because of how the indexicals are used. The substantivalist could also simply accept that there are distinct yet indistinguishable possibilities.

[13] The substantivalist can agree that the "hole diffeomorphism" does not yield a distinct situation, by denying that mere haecceitistic differences amount to physically distinct situations, in particular because of the symmetry in the physics: Brighouse (1994). This is a popular response (which has come to be called "sophisticated substantivalism"), though there are dissenters (such as Belot and Earman (2001)). Hoefer (1996); Dasgupta (2009, 2011); Baker (2010); Arntzenius (2012, Sec. 5.12) also argue that symmetries support this kind of reply.

(Above I said that the facts or quantities recognized by the laws, given in part by the laws' invariances, indicate which worlds or situations count as distinct according to a given theory. However, the static shift reveals that the (non)invariances of the laws and phenomena needn't on their own entail the distinctness or not of the shifted situation. Thus, you might think, and it is often said, that spatial homogeneity is needed to generate the static shift argument. But a non-homogeneity, such as a preferred location picked out by the laws, would not immediately entail that the static shift generates a distinct situation. That depends on how we identify the preferred location in the shifted case, for example whether the preferred location gets "carried along" with the shift, as well as one's views on haecceitism and related theses (it depends on what counts as qualitative similarity between worlds, for instance). That said, spatial homogeneity is required for Leibniz's own arguments.)

One final example I will briefly mention is from Kant. Kant argued that the relationalist is unable to say whether a glove in an otherwise empty world is left- or right-handed. The relationalist facts, the spatial separations between corresponding parts of the glove, are the same either way. Yet surely, Kant suggests, there is a fact about whether a lone glove is left- or right-handed.

However, it is first of all open to the relationalist—as well as the substantivalist, for that matter—to deny that there is a fact of the matter, in an otherwise empty world, as to *which* handedness a glove has, even while allowing that there is a fact that it is *handed*, or chiral. (A handed or chiral object is one that cannot be mapped onto its mirror image by means of rigid rotations and translations. It is identical in all respects to its mirror image aside from a mirror reflection; for example, a glove in three-dimensional space or the letter *F* in a two-dimensional plane.[14]) A lone glove is asymmetric in this way. That it is chiral is plausibly intrinsic to the glove, having to do with the geometrical relations among its parts, even though, in an otherwise empty world, there is no fact about whether it is left-handed or right-handed. Second, handed objects, and parity-violating behavior more generally, simply show that the relationalist's physics may be surprisingly non-local. Odd, but I believe not fatal to the view.[15]

I take these examples to be inconclusive to the traditional question of the existence of space. Each one aims to show that the opposing side posits either

[14] This idea assumes that the space the object inhabits is orientable: *F* does not differ in the above way from its mirror image in a two-dimensional Möbius strip, for instance. This may lead you to wonder whether the relationalist can account for handed objects after all, since there must be an account of the background spatial structure required for there to be this kind of object. Brighouse (1999) argues that the relationalist can account for this by adopting a modal relationalism (Section 5.5 below).

[15] See the discussion in Arntzenius (2012, Sec. 5.4), who sees it as near-fatal. Albert points out (in a seminar at Rutgers in Spring 2020) that the resulting non-locality is different in kind from that of quantum mechanics. Here, the non-locality stems from the nature of the fundamental quantities used to define systems' states, which are themselves non-local—relative rather than absolute positions, for example. This is a kinematical non-locality, not the dynamical non-locality we find in quantum mechanics, and is arguably less troubling.

too few or too many spatial facts for the physics, with the relationalist facing the charge of positing too few and the substantivalist the charge of positing too many. What we have seen is that each side seems able to adequately respond to the challenge, by claiming that it does posit all the right spatial (or spatiotemporal) facts. To the extent that we wish to rely on these cases to settle the dispute, as was traditionally done, we have reached a stalemate. (Nor does the assessment change in view of more updated physics: note 10.) Since these cases continue to be a focus of discussion even now, it is no wonder that many people feel this dispute is at a stalemate or has become irrelevant to ordinary physics, turning on such things as the metaphysics of modality. At this stage of things, one might well wonder "why anyone, other than a few academic philosophers with a bent for far-side metaphysics, should care about the tug of war over whether space-time is a substance" (Earman, 1989, 163).[16]

5.3 A lesson of the traditional examples

There is nonetheless an important lesson to be had from thinking through the traditional examples and various responses to them, which is this: the relationalist and substantivalist can both account for the phenomena so long as they posit enough, but not too many, spatiotemporal facts for the physics. Whatever your view on the ontology, you should be careful to posit the right spatiotemporal facts or structure for the physics. Of course, the substantivalist and relationalist may disagree on which facts must be recognized by a proper physics. But cases like the ones above suggest certain constraints on those facts, depending on the physics in question—as the bucket and globes suggest that facts about rotation should be recognized by any viable Newtonian physics. (From now on I will be discussing things explicitly in spacetime terms.)

Although we reached that conclusion by thinking about those particular examples, a similar conclusion follows from the discussion in earlier chapters. In Section 5.2 we saw that in the face of the bucket and globes, the relationalist should posit enough spatiotemporal relations to undergird facts about rotation, and in the face of the kinematic shift, the substantivalist should deny that there are facts about absolute velocity. Both of these conclusions are captured by the claim that one should posit a Galilean spatiotemporal structure, which supports facts about acceleration but not about absolute velocity—the same conclusion we reached in Chapter 3. Both the particular examples above, and Newton's laws in general,

[16] Earman goes on to suggest that the hole argument breathes new life into the debate. However, the literature since then has made it clear that the hole argument, too, turns on metaphysical theses (in particular concerning modality and determinism) that are independent of relationalism versus substantivalism per se.

indicate the need for an inertial structure, but not an absolute space structure. In this way, the traditional examples serve to highlight something we saw earlier: the physics, assumed here to be Newtonian, requires the structure to distinguish between inertial and non-inertial motions, but not between motions with different constant velocities.

Put another way, the traditional examples indicate that the relationalist and substantivalist should both adhere to the matching principle mentioned in Chapter 3; for doing so allows each side to claim the most plausible responses to the cases leveled against it. Taken together, the traditional examples suggest that, regardless of whether you are a relationalist or a substantivalist, you should impute to the world the structure or facts required for the physics, and you should not impute more structure than that. Assuming a Newtonian physics, this means that you should attribute a Galilean spatiotemporal structure to the world.

To say that the relationalist and substantivalist both *should* posit this structure or those facts is not yet to say that they both *can* do this, and you may wonder whether the relationalist in particular can do this. How can the relationalist believe in spatiotemporal structure, let alone posit the "right" such structure, if there is no such thing as spacetime? Alternatively, you might worry that if the relationalist can somehow do this, then the view will become so close to substantivalism as to eradicate any real difference between the two. I turn to this in Section 5.4, where I discuss how to make sense of the idea that each view can believe in spatiotemporal structure, and even posit the right such structure, all the while retaining a substantive disagreement between them.

First let me note a few other positions one could take. The matching principle assumes that there is such a thing as the spatiotemporal structure of a world, which we learn about from its physics. This assumes a realism about spatiotemporal structure. An opposing view is conventionalism. Reichenbach (1958) argues that there is no nonconventional sense to be made of "the" spatiotemporal structure of a world. He notes that our empirical evidence, the various spatial and temporal relations we measure between material bodies, will underdetermine the spatiotemporal geometry. Suppose that our measurements seem to indicate a curved spatial region; say, we measure the interior angles of triangles to sum to more than 180°. Reichenbach points out that this evidence is compatible with the space's being everywhere flat and Euclidean, for there could be undetectable universal forces, which distort all material things, including our measuring instruments, in uniform, hence undetectable, ways.

Whether we conclude that space is curved will then depend on whether we countenance universal forces. And since we cannot empirically determine whether there are such forces, Reichenbach concludes that whether we allow for universal forces, and at what strength, is a conventional choice to be made on the basis of pragmatic considerations. Relative to such a choice, we may say that the world has a particular spatiotemporal geometry; but there is no fact of the matter about the structure independent of that choice. The choice concerning universal forces, and thus the spatiotemporal structure that depends on it, is "an arbitrary decision that is neither true nor false" (Reichenbach, 1958, 19). No one choice is objectively more correct than any other.

I do not have a conclusive argument against conventionalism about spatiotemporal geometry. I have been assuming a realism that effectively rejects it out of hand. I have been assuming that there is such a thing as the spatiotemporal structure of a world, which we aim to discover in part by relying on certain epistemic principles, and which we can be right or wrong about. That said, notice how the discussion in Section 5.2 lends further support to this outlook. In response to the traditional examples from the debate over the ontology of space, we were led to say that one side or the other should posit the right spatiotemporal structure for the physics, a conclusion that assumes there is an objective, determinate fact about a world's spatiotemporal structure. The traditional examples push us toward a realism about spatiotemporal structure, and in a way that is independent of whether one is a relationalist or substantivalist. (We can agree with Reichenbach that certain choices—for example of unit of measure or coordinate origin or inertial reference frame—are arbitrary. That is because the physics tells us that different such choices yield equally good descriptions: the laws remain the same regardless of which choice we make. Spatiotemporal structure is different. We cannot arbitrarily alter the spacetime metric, for instance, while keeping the laws the same.)

Another kind of view denies that there is such a thing as a spatiotemporal structure that is presupposed by the laws, and which can come apart from them. Albert (2019b, 2020) takes the Reichenbach-style examples to suggest that the dynamics does all the work in generating the appearances of a spatiotemporal geometry. He concludes that there is no such thing as a world's pre-dynamical spatiotemporal structure. Rather, claims about spacetime and its structure are just ways of saying things about the dynamical laws.[17] (A version of this idea is held in the dynamical approach to spacetime of Brown (2005), Brown and Pooley (2006); the related approach of Myrvold (2019) discussed in Section 3.3; and the spacetime functionalism of Knox (2013, 2019).)

I do not have a conclusive argument against this kind of view either. I simply think that we should draw a different conclusion from Reichenbach's examples. Although the dynamics plays a role in producing the geometrical appearances, we should not go so far as to conclude that there is no fundamental spatiotemporal structure. Rather, we should conclude that the dynamics gives us (fallible) *evidence*

[17] Albert allows for fundamental differentiable and topological structure, but nothing beyond that, and so nothing that intuitively counts as spatiotemporal. As Baker puts it, that would be "quite an impoverished conception of spacetime" (2005, 1303), lacking most of the features we usually take to be essentially spatiotemporal (as argued by Maudlin (1988); Hoefer (1996)).

for what the underlying spatiotemporal structure *is*. Recall from Section 3.3 that Albert's view must reject certain familiar inferences from physics, which assume that the laws and spatiotemporal structure can come apart; for example, that Lorentz's ether theory posits a different spatiotemporal structure from Einstein's theory of special relativity, and one that is less preferable, given the laws. For Albert, Einstein's and Lorentz's theories are at bottom the same theory, since they have the same dynamics: there is no genuine choice to be made between them. More generally, as mentioned earlier, a view like Albert's simply rejects one of my starting points, the basic thought that the laws require or presuppose some spatiotemporal structure in order to be meaningfully formulated, and which we therefore ought to posit.

Another type of view is spacetime structural realism.[18] This view starts from a different notion of spacetime structure than I do, one that is in line with the alternative conception of structure adopted in the structural realism literature. My own approach doesn't say anything about the relative fundamentality of objects (or intrinsic properties) as opposed to structure, which is the focus of much of the spacetime structural realism literature in particular, and the structural realism literature in general. In particular, I do not advocate demoting the metaphysical status of objects (or intrinsic properties) as opposed to relations. Spacetime structural realism furthermore claims to be a third view, distinct from both relationalism and substantivalism, whereas I suggest that the relationalist and substantivalist should themselves both be realists about spatiotemporal structure. In all these ways, I simply have a different approach to and conception of a realism about spatiotemporal structure.

5.4 A disagreement about ground

So far I have suggested that both the relationalist and the substantivalist should countenance or believe in spatiotemporal structure, in order to be able to posit the right such structure for the physics, as indicated by the traditional examples discussed in Section 5.2 as well as the guiding principles from Chapter 3. This raises an obvious question: *can* they both believe in this?

In this section I argue that they can. I argue that both the relationalist and the substantivalist can countenance or believe in the existence of spatiotemporal structure. (Whether each one can countenance the particular structure for the physics is a question I will be sidestepping, for reasons to come.) Furthermore, the way that they each can do this will give rise to a genuine disagreement between them, one that is relevant to physics.

[18] See for example Dorato (2000, 2008); Slowik (2005); Bain (2006); Esfeld and Lam (2008); Ladyman and Ross (2009); see also Greaves (2011).

The basic idea is this. (I outline the main idea here, postponing further details until Section 5.5.) Each view can countenance or believe in the existence of objective, determinate spatiotemporal structure or facts. (These facts are structural in the sense of Chapter 2. They concern the intrinsic, genuine, objective spatiotemporal features of a world, those that don't depend on arbitrary or conventional choices in description—for example, that two objects are separated by some distance under a Euclidean metric, or that a certain particle's trajectory is straight according to a given inertial structure.) The disagreement between the views concerns what underlies this structure or those facts. The substantivalist says that spatiotemporal structure is fundamental to the physical world. The relationalist says that spatiotemporal structure is not fundamental, but instead arises from the relations between and properties of material bodies.

I am now going to spell out this disagreement in terms of a grounding relation; though what is most important is to make use of some notion of relative fundamentality. For reasons that will become clear, I suspect that ground is best suited to the job, but you may substitute some other such relation if you prefer.

A grounding relation is an explanatory relation that captures the way in which one thing depends on or holds in virtue of another. Ground captures a metaphysical *because* in answer to questions about why something exists or some fact holds. Importantly, the holding of a grounding relation between two objects or facts does not imply the non-existence of the grounded object or fact. (Philosophers disagree about whether ground holds primarily between objects or facts. I wish to remain neutral on this, and will alternate between both conceptions. I also aim to remain neutral on surrounding issues about the metaphysics of ground. I will assume that the grounding relation is transitive and irreflexive, and that the grounds metaphysically necessitate the grounded, assumptions that are reasonably standard, if not wholly uncontroversial.[19])

Recall the traditional debate, put into spacetime terms. The substantivalist claims that spacetime exists: the physical world comprises, in addition to material bodies, an independently existing space(time), and facts about the spatiotemporal relations among material bodies have to do with where these things are located in this space. The relationalist denies that spacetime exists. The physical world comprises only material bodies with various properties and relations between them, including spatiotemporal ones. There is no additional "container" in which material bodies are located.

Now put these ideas in terms of ground. The relationalist says that a world's spatiotemporal structure is grounded in, it holds in virtue of, the features and behavior of material bodies. All the spatiotemporal facts are grounded in facts about material bodies. The substantivalist denies this. Spatiotemporal structure

[19] Different accounts of ground are in Fine (2001, 2012); Schaffer (2009). Rosen (2010) defends the general idea.

is not grounded in anything else more fundamental to the physical world; in particular, it does not hold in virtue of anything having to do with material bodies.

According to this way of conceptualizing the two views, the substantivalist and the relationalist can agree that spatiotemporal structure exists. They can both say that there are objective spatiotemporal facts about a world. They disagree about what, if anything, grounds this structure or those facts. They disagree about what the spatiotemporal structure holds in virtue of; what metaphysically explains the spatiotemporal facts.

Thus, suppose that a certain world of classical point-particles has a Euclidean spatial structure, or that a Newtonian world has a Galilean spatiotemporal structure. And suppose the relationalist and substantivalist agree that the given world has the stated structure. How can each view make sense of this?

Start with the relationalist. The relationalist believes that the fact that a world has a certain spatiotemporal structure is made true by the facts about material bodies; a world has the spatiotemporal structure it does because material bodies behave in certain ways. A given world then has a Euclidean spatial structure *because*—in the metaphysical sense—its particles move around in various ways, with various spatial relations between them, which conform to the postulates of Euclidean geometry. The fact that the world has a Euclidean spatial structure is grounded in, holds in virtue of, the fact that its particles are, and can be, arranged in those ways. (I return to this "can be" phrase in Section 5.5.) Likewise, the fact that a Newtonian world has a Galilean spatiotemporal structure is grounded in the fact that its particles do, and can, behave in various ways, with various spatiotemporal relations between them. The world's spatiotemporal structure is Galilean *because* (in the metaphysical sense) its particles behave in certain ways. That's all there is to the world's having the stated structure.

It is worth emphasizing that ground yields "a distinctive kind of metaphysical explanation," in Kit Fine's words, in which the objects or facts are connected by a "constitutive form of determination" (2012, 37). Particle behaviors don't cause a Euclidean spatial structure, for instance; rather, this is what the spatial structure *consists in* or depends on in a metaphysical sense. Compare the grounding of facts about the macroscopic world in facts about subatomic particles, or the grounding of mental facts in non-mental facts, or moral facts in non-moral ones. Ground captures this kind of metaphysical "in virtue of" explanation.[20] To say that "the fact that *x* grounds the fact that *y*" just means that "the fact that *y* holds in virtue of the fact that *x*," that the holding of the grounded fact consists in nothing more than the holding of the grounding fact.

The substantivalist, by contrast, denies that a world has a given spatiotemporal structure in virtue of the behavior of its material bodies. It is a fundamental fact

[20] See Loewer (2001) on the relevant sense of "in virtue of."

that a given world has a Euclidean spatial structure, and it is in virtue of this fact that various spatial relations hold among the particles. Thus, the fact that two particles are a certain distance apart is grounded in, made true by, the fact that the particles are separated by that amount according to the fundamental metric structure. (The metric structure may be understood in different ways by different substantivalists, as I discuss in Section 5.5, but in any case will not be grounded in features of the particles.) It is likewise a fundamental fact about a Newtonian world that it has a Galilean spatiotemporal structure. A given particle then follows a straight trajectory because (in the metaphysical sense) its path is straight according to the fundamental inertial structure.

In this way, both the substantivalist and the relationalist can countenance spatiotemporal structure or facts; they disagree on what, if anything, grounds that structure or those facts. Compare Fine's (2012, 42) idea that the notion of ground clarifies the debate between three- and four-dimensionalists about material objects. According to Fine, we should not say that the difference between these views lies in whether one believes in the existence of temporal parts, since even the three-dimensionalist can allow that there are such things as temporal segments of material objects. Instead, the difference lies in the question of ground: in whether the existence of a temporal part at a time is grounded in the existence of the persisting object at that time, as the three-dimensionalist will say and the four-dimensionalist will deny. Analogously, the relationalist and the substantivalist can both believe in the existence of spatiotemporal structure. The difference lies in the question of ground: in whether the existence of a spatiotemporal structure is grounded in the features and behavior of material bodies, as the relationalist will say and the substantivalist will deny.

(The analogy to the debate between three- and four-dimensionalists could give fodder to certain philosophers who feel that the spacetime debate is not substantive. If you do not think that the former is a real disagreement, then substitute some other dispute that you do think is substantive, for which the opposing sides agree on the existence of certain objects or facts, but disagree on what, if anything, metaphysically explains those objects or facts. Consider the debate in the metaphysics of quantum mechanics between those who say that ordinary three-dimensional space is fundamental and those who say that it emerges from a more fundamental, extremely high-dimensional space. This can be seen as the question of ground: a debate about what, if anything, grounds the existence of, or facts about, three-dimensional space.[21])

You might think that the debate, put in these terms, cannot be a genuine dispute, for the following reason. If the relationalist says that spatiotemporal structure is nothing but the spatiotemporal relations among material bodies, and since

[21] I put the debate in these terms in North (2013).

material bodies (and various of their properties and relations) are fundamental, doesn't this just amount to the view that spatiotemporal structure is itself fundamental—which is exactly what the substantivalist claims?

This is where the notion of ground comes in. Compare the reductionist who says that the macroscopic facts about physical systems are distinct from, yet are also nothing over and above, the microscopic facts about systems' particles. This is prima facie puzzling. How can one thing be *both* different from *and* nothing but another? Ground is useful here precisely because it allows us to capture both these thoughts. The reductionist can say that the facts about the macroscopic world are grounded in the facts about subatomic particles—which is to say that there are real facts about macroscopic physical systems, which are *distinct* from the facts about their particles, but are *also* nothing over and above the facts about the particles. Exactly how, and whether, the notion of ground can accomplish this two-headed task is a big question. For our purposes, I am going to assume that this is exactly what the notion is designed to do, and leave it to the experts to work out how it can be done.

In saying that facts about spatiotemporal structure are grounded in facts about material bodies, the relationalist likewise says that there are facts about a world's spatiotemporal structure, which are distinct from the facts about material bodies and their relations, yet are also nothing over and above those facts about material bodies. It may seem as though the spatiotemporal structure just is those relations, but this feeling simply comes from the fact that "there is no stricter or fuller account of that in virtue of which the explanandum holds" (Fine, 2012, 39). By contrast, the substantivalist says there are facts about spatiotemporal structure that are not grounded in facts about material bodies, which are in that way over and above any facts about material bodies.

If someone were to ask, then, "Why (in the metaphysical sense) does the world have a Galilean spatiotemporal structure?", the relationalist will say: "because the particles behave thus and so"; this is the "strictest or fullest account" to be had. The substantivalist will have no answer: the fact that a world has a Galilean structure is a fundamental fact, already the strictest or fullest account to be had. (More exactly, there may be an answer, depending on the version of the view, but it will not reference material bodies: Section 5.5.)

Substantivalism and relationalism, as I see them, *disagree about the fundamental nature of the physical world.* They both countenance spatiotemporal structure or facts, but disagree on whether all such structure or facts hold in virtue of material bodies. Both views can recognize the fact that two particles are separated by some distance under a Euclidean metric, or that a given world has a Euclidean metric structure; they will disagree on whether the metric is itself fundamental or grounded in the features and behavior of material bodies. This is a real disagreement: a substantive debate about what makes it the case that the spatiotemporal structure needed for the physics holds.

Although this conception of the debate is not quite the traditional one, it captures the core ideas behind that dispute. Dainton notes that,

> Substantivalism, in its traditional guise, is the doctrine that space is an entity in its own right, possessing its own inherent structure. The existence and structure of such a space is (to some degree at least) independent of the existence, distribution and behaviour of any material bodies that it may happen to contain.
>
> (Dainton, 2010, 250)

On my conception of the dispute, these thoughts are captured by the claim that the substantivalist believes that (facts about) spatiotemporal structure are fundamental, in particular not grounded in (facts about) material bodies, and which are in that way independent of, or over and above, anything having to do with material bodies. By contrast, the relationalist maintains that material bodies and their features metaphysically explain a world's spatiotemporal structure, and in this way the "existence and structure of space" is not independent of material bodies. The relationalist can still say that spatiotemporal structure exists, there are objective truths about what spatiotemporal structure a world has—it's just that these things all hold in virtue of what's true of material bodies.

I spell out some further details in Section 5.5, but first a few notes on this conception of the debate compared to some others in the literature.[22] In contemporary discussions, one can find the thought that the relationalist believes in the existence of "spacetime," understood as being somehow constructed out of material bodies and their features. So it may seem as though the traditional dispute itself (as well as contemporary takes on it) was never about the existence of spacetime but its fundamentality, and my own formulation may seem like merely a new label for an old dispute.

However, that is something of an anachronism. Traditional participants, like Newton and Leibniz, were not thinking explicitly in terms of fundamentality. Neither, of course, were they thinking in spatiotemporal terms. At the same time, to the extent that we can understand what they were saying in these terms, this reveals that my understanding of the debate is, as I claim, an updating of the traditional dispute, taking into account more recent developments in physics (involving spacetime and its structures) and philosophy (fundamentality and ground).

Jonathan Schaffer (2009, 363) and Dasgupta (2011) also suggest that we construe the spacetime ontology debate in terms of ground. They say that the relationalist and substantivalist both believe that spacetime exists, while differing on what grounds the existence of spacetime. I say that the relationalist and substantivalist both (can and should) believe that spatiotemporal structure exists, while differing on what grounds the existence of this structure. These conceptions of the dispute

[22] Additional comparisons are in North (2018, Sec. 3.4).

are similar, for many intents and purposes the same. The difference is that my way of putting things allows us to spell out the relationalist and substantivalist viewpoints in a variety of ways, corresponding to various conceptions that have been put forth in the literature, while still maintaining a substantive dispute between them, as we will see at the end of Section 5.5.

It may seem unexciting to exchange a debate about the existence of spacetime for one about the fundamentality of spatiotemporal structure. Recent discussions in metaphysics have proposed making a similar exchange with various debates about existence questions for debates about fundamentality (as in Schaffer (2009)). There have been some related thoughts in recent philosophy of physics as well. Carl Hoefer frames the question about spacetime as one of "*trying to understand the basic ontology of the physical world*" (1998, 466; original italics), that is, of what's fundamental to the physical world. Hoefer, too, argues that this is a substantive dispute, which is likely to remain so with future physics, and that general relativity supports substantivalism (as I will suggest in Section 5.6). However, he formulates aspects of the dispute more traditionally, saying for instance that substantivalism is committed to the existence of "a substantial, quasi-absolute entity" (Hoefer 1998, 464), and he reaches those conclusions for different reasons.

The approach of Belot (1999, 2000, 2011) is perhaps closest to my own. Belot says that the relationalist can be a realist in the sense of "attribut[ing] to reality a determinate spatial structure," while disagreeing with the substantivalist over "the nature of the existence of space" (2011, 1–2). Relationalism and substantivalism are then

> attitudes which one can adopt towards physical geometry in general. We can agree about the geometrical structure of space or of spacetime, and about the constraints that this imposes upon the actual and possible relations between material bodies and events—and then go on to give either a substantivalist or a relationalist reading of this geometry. (Belot, 1999, 36)

Translating this into my terms: the relationalist and substantivalist can both believe in spatiotemporal structure; they can even posit the same spatiotemporal structure to a world; they disagree about what gives rise to that structure. Belot, too, says that his conception of the dispute, though unorthodox, yields a debate that is substantive, relevant to physics, and reminiscent of the traditional dispute. However, his account is not spelled out in the same way—it does not make use of notions like ground or my conception of spatiotemporal structure, and it remains focused on some of the traditional examples—and he does not draw the same conclusions. (He comes out in favor of relationalism.) That said, his basic conception of the dispute is in the same spirit as my own.

In any case, the more prevalent attitude in philosophy of physics, especially among those who complain about the substantivity of the dispute, is that the debate concerns the existence question. So although my understanding of the dispute is not without precedent, even then there are differences, and it is in any case not the prevalent viewpoint. If you still disagree with that assessment, however, it will soon become clear that novel avenues of argument open up once we are completely explicit about this shift.

5.5 The disagreement: further details

Above I outlined the gist of the dispute as I see it. In this section I discuss some further details, complications, and clarifications.

The relationalist says that certain material bodies, and certain of their properties and relations, are fundamental, and a world's spatiotemporal structure holds in virtue of them. (The statement that "certain material bodies are fundamental" is meant to refer to whichever material things turn out to be most fundamental, perhaps certain kinds of material particles.) All spatiotemporal structure or facts are grounded in (facts about) material bodies. (I assume that the fundamental relations can include spatiotemporal ones, although the relationalist might prefer a different type of relation to be fundamental, causal ones being a familiar candidate. I leave this open here.[23])

At this point you might wonder how the facts the relationalist takes to be fundamental manage to ground all the spatiotemporal facts needed for the physics. Merely being a realist about spatiotemporal structure does not guarantee the ability to generate the particular structure that's required (as the matching principle demands). Much of the literature is taken up with this question of how, and whether, the relationalist's more meager ontology can recognize all the spatiotemporal facts we want. And you may be skeptical that the relationalist can do this. A repeated complaint against the varieties of relationalism surveyed by Pooley (2013), for example, is that the relationalist's resources are too thin to yield the right predictions.

Perhaps surprisingly, I won't try to tell you how the relationalist grounds all the requisite spatiotemporal facts in facts about material bodies.[24] In Section 5.6, I am going to propose an argument for substantivalism that goes through even if we grant the relationalist the ability to ground all the relevant facts in those taken to be fundamental. So for the purposes of that argument, I am going to grant the relationalist that ability.

That said, let me mention one thing that I suspect will be required, which is some version of "modal relationalism." Modal relationalism countenances facts

[23] See Nerlich (1994, Ch. 1) for argument that the relationalist's fundamental relations cannot be spatiotemporal. I further assume that objects and relations are equally fundamental, though there might be a version of the view with only one fundamental ontological category, in the sense of Paul (2013).

[24] From this perspective, those such as Manders (1982); Mundy (1983, 1992); Huggett (2006); Belot (2011) can be seen as giving accounts of how this grounding project might go.

not only about the actual features and behavior of material bodies, but about their possible ones as well—facts about what spatiotemporal relations *can* hold, in some sense. The reason to think this will be required is the following. So long as one can embed the actually instantiated relations uniquely into a certain structure, it seems as though the relationalist can talk of the spatiotemporal structure of a world. However, the actual relations may not uniquely fix the structure (up to isomorphism) needed for making predictions about the behavior of material bodies.[25] In order to adhere to the matching principle, in other words, the relationalist will have to go modal. See Brighouse (1999) and Belot (2011) on how the relationalist might do this and what type of modality may be involved.

Notice the relationalist might not deny the fundamentality of *any* spatiotemporal facts or structure. Depending on the version of the view, the fundamental facts may include ones such as that two particles are separated by some distance, or that one event happens earlier than another—facts that are spatiotemporal in nature. What's important is that the relationalist only allows for certain kinds of spatiotemporal facts (if any) to be fundamental, namely those that essentially involve material bodies and their relations—facts the substantivalist takes to be nonfundamental. For the relationalist, the fact that a world has a certain spatiotemporal structure is grounded in facts about material bodies, even though these latter facts may include certain spatiotemporal ones. More exactly: there is no fundamental spatiotemporal fact or structure apart from the structure of, or facts about, material bodies. (Since some spatiotemporal facts or structure may be fundamental, hence ungrounded, assuming that fundamental facts are ungrounded.) For ease of exposition, I put this as the claim that all spatiotemporal facts are grounded in facts about material bodies; all spatiotemporal structure is grounded in the relations between and properties of material bodies.

A related precisification holds of substantivalism. The substantivalist denies that all spatiotemporal structure or facts hold in virtue of (facts about) material bodies. A world's spatiotemporal structure is not grounded in the features and behavior of material bodies. The fact that a world has a certain spatiotemporal structure is a fundamental fact, which in turn grounds the facts about the spatiotemporal relations between material bodies. (The former may only partially ground the latter, since the grounds may include occupation relations that material bodies bear to spacetime points or regions, depending on the version of the view: more at the end of this section.) The substantivalist thus recognizes nonfundamental spatiotemporal facts or structure of a sort, about the spatiotemporal relations between material bodies. More exactly, the view holds that there are fundamental spatiotemporal facts or structure not grounded in (facts about) material bodies. (Certain facts about material bodies, for instance about their fundamental intrinsic

[25] Examples are in Mundy (1986); Maudlin (1993, 193–94, 199–200); Nerlich (1994); Belot (2000, 2011, Ch. 2). Modal relationalism arguably even allows the view to accept the possibility of vacuum worlds, often seen as particularly problematic for the relationalist.

properties, will be fundamental. What's not fundamental are the spatiotemporal facts about them.)

You might think this conception of substantivalism is already disconfirmed by our current best theory of spacetime. According to general relativity, the presence of matter affects the local spatiotemporal geometry, which in turn affects the behavior of matter. Yet according to my conception of substantivalism, there is spatiotemporal structure that is independent of matter.

The thought is misplaced; for the inter-dependence between spatiotemporal structure and material bodies in general relativity is of a different, causal or nomological, kind from that given by ground. Although the substantivalist says there is spatiotemporal structure that is independent of material bodies in the sense of not being grounded in them—facts about spatiotemporal structure are metaphysically over and above facts about material bodies—the substantivalist can still allow that the behavior of material bodies causes a certain spatiotemporal structure in accordance with the physical laws. Compare: although the dualist thinks that mental events are not grounded in physical events—mental events are metaphysically over and above physical ones—the dualist can still allow that physical events cause mental events in accordance with the scientific laws.

Now, the substantivalist might not take a world's spatiotemporal structure to be absolutely fundamental. Newton held that absolute space is a necessary consequence of God's existence. Suppose he also held that absolute space is less fundamental than God. (It is not clear that he would say this, but this could be a way of spelling out the idea.) Newton would still count as a substantivalist, as I conceive of the view, because the facts about the spatial structure are more fundamental than the facts about bodies' spatial relations. To put it another way, facts about the world's spatial structure are fundamental *to the physical realm*. Relatedly, the relationalist denies that such facts are fundamental to the physical realm, regardless of whether there is alleged to be something yet-more-fundamental outside the physical realm altogether. The two views still disagree about whether spatiotemporal structure apart from material bodies is fundamental to the physical world. For ease of presentation, I will continue to put this as a disagreement over whether spatiotemporal structure is fundamental (to the physical realm).[26]

What if there is no fundamental level to the physical realm? In that case, the views may still disagree on the relative fundamentality of material bodies and a world's spatiotemporal structure, depending on the details. This may seem to suggest that the debate should be framed in the following way: substantivalism

[26] Although I think it most natural to see my view as characterizing Newton as a substantivalist (and Leibniz as a relationalist), there is room for debate. I won't take a firm stand on how best to construe those views in my terms (even though I do claim to be capturing the core of what historical disputants were getting at).

maintains that the facts about a world's spatiotemporal structure are more fundamental than the spatiotemporal facts about material bodies; relationalism maintains that the facts about material bodies are more fundamental than the facts about spatiotemporal structure. However, that conception of the dispute would imply that one of either relationalism or substantivalism is bound to be correct, regardless of future physics, so long as the two kinds of facts are not equally fundamental. But if nothing like either spatiotemporal structure or material bodies turns out to be fundamental to the physical world, then it seems as though neither view has been vindicated. One could insist that substantivalism would still be correct so long as the facts about the world's spatiotemporal structure are more fundamental than the spatiotemporal facts about material bodies, and contrariwise for relationalism. This strikes me as too far removed from the original ideas. More importantly, I don't think that one of these views must be correct regardless of future physics, and it will depend on the details of that future physics whether one or the other—or neither—is correct.

My conception of the dispute allows us to sidestep many of the reasons people feel the traditional dispute has stagnated or become non-substantive. It does not center on the traditional cases and particular replies to them, nor on how to count possibilities or one's metaphysics of modality more broadly, things that are detachable from the spacetime debate (a point emphasized by Dasgupta (2011)). It also avoids having to draw certain distinctions that many people have been skeptical of, such as between container and contained, substance and non-substance, absolute and relative.[27]

There are a few distinctions presupposed by my understanding of the dispute, but they are not as unclear as those required by more traditional conceptions. First, it assumes a distinction between the fundamental and the nonfundamental, which is something we have a reasonably clear pre-theoretic grasp of. Second, it requires that we can pinpoint the structures that count as spatiotemporal as opposed to those that don't. This is something the physical laws give us a handle on, in ways discussed earlier, though more could be said. Perhaps there is nothing else that makes a fact or structure spatiotemporal; perhaps there is.[28] Either way, I take the idea to be familiar enough from physics, which provides us with some clear cases. Third, my conception requires a distinction between material bodies and other things in the world. Although people have worried about the clarity of this distinction,[29] I think it is clear enough for our purposes. At the least, I suggest that we understand the debate in this way, on the assumption that we will be able to locate such a distinction. In this I follow Earman, who says that, "it is a delicate and difficult task to separate the object fields into those that characterize the space-time structure and those that characterize its physical contents," while noting that "the

[27] Rynasiewicz (1996, 2000) worries about the clarity of all of these and more besides.

[28] Belot (2011); Brighouse (2014); Knox (2019); Baker (2020) are different accounts.

[29] See especially Rynasiewicz (1996).

vagaries of this general problem need not detain us here, since there are enough clear cases for our purposes" (1989, 155–6).[30]

Most importantly, this characterization of the dispute leaves room for future physics to provide an answer, so that this dispute cannot be "merely verbal" or "purely metaphysical." We think there is a real difference between a world in which spatiotemporal structure is fundamental, and a world in which it arises from some pre-spatiotemporal structure, for instance. Physicists treat these as genuinely different possibilities, governed by different theories. This may be clearest in the case of different approaches to quantum gravity, many but not all of which appear to treat spatiotemporal structure as arising from something fundamentally non-spatiotemporal (more at the end of Section 5.6). This is evidence of a genuine difference between the views, as I see them.

There is another way of framing the dispute that is familiar and seems close to what I have said: according to the substantivalist, there exists a fundamental physical space(time); according to the relationalist, there does not. Similarly, the relationalist denies, whereas the substantivalist accepts, the existence of spacetime points or regions as fundamental physical objects. The problem with this way of putting things is that it is not always clear what it means to say that a physical space or spacetime points exist. I suspect that something like this is the main reason behind the feeling that the debate is unclear, especially within the philosophy of physics community. Myrvold (2019, 139) says that the idea of a substantival space is no more than a metaphor based on unfounded intuition. Other philosophers of physics worry about taking spacetime points to be concrete physical entities in particular; as Malament says, "They certainly are not concrete physical objects in any straight-forward sense. They do not have a mass-energy content.... They do not suffer change. It is not even clear in what sense they exist *in* space and time" (1982, 532). Others have worried more generally that this kind of ontological dispute—a dispute that is just about what things exist—is non-substantive or merely verbal.[31] Howard Stein says that, "the word 'ontological' itself presents seriously problematic aspects," in particular, "Quine's usage [is] not a very useful one for the philosophy of physics" (1977a, 375).

As I see it, the debate is about the fundamentality of spatiotemporal structure, in particular whether there is any spatiotemporal structure or fact not grounded in the features of, or facts about, material bodies. Within this framework, there is some flexibility as to how to formulate the dispute. Neither the matching principle nor my conception of structure says how we must construe the nature of spatiotemporal structure; nor have I taken a stand on whether to construe

[30] See Hoefer (1998) for argument that the distinction can generally be made.

[31] This seems the spirit behind Stein (1970a, 1977a) and Curiel (2018), perhaps also Wallace (2012). There have been similar thoughts expressed in metaphysics, for example by Hirsch (2011), but it is not clear that it is exactly the same idea as what's being expressed by philosophers of physics.

ground as a relation between objects or facts. As a result, although we can put the disagreement as being about whether there exists a fundamental physical spacetime or fundamental spacetime points, we do not have to. Those who are squeamish about putting things in ontological terms can still see the debate as being about the fundamentality of spatiotemporal structure, understanding this as being a dispute not about whether there exist certain objects (over and above material bodies), but about whether there are certain facts (over and above the facts about material bodies). On this conception, the relationalist says that the fact that a world has a certain spatiotemporal structure holds in virtue of the fact that material bodies behave thus and so, and the substantivalist denies this, seeing it as a fundamental fact about the physical world. This allows us to discuss the dispute, and evaluate the evidence for either side, while remaining neutral on how the substantivalist wants to understand the instantiation of that structure or the ontology behind that fact.

This dovetails with an idea in the spacetime structural realism literature. Jonathan Bain (2006) argues that classical field theory (this includes general relativity), standardly given in terms of a tensor formalism, can be mathematically formulated in ways that do not presuppose a differentiable manifold of points. He describes three alternative formalisms one could use—twistor theory, Einstein algebras, and geometric algebra—none of which treat points as fundamental. My understanding leaves it open for the substantivalist to spell out the spatiotemporal structure in any of these ways, or even to refuse to choose among them, as Bain himself proposes. (Bain concludes that we should be realists about spacetime structure but not about any particular instantiation of it. He sees this as a third view, since according to him the substantivalist is committed to spacetime points, but it counts as substantivalist by my lights.[32])

To be explicit, my conception of substantivalism can encompass a few different kinds of view, each of which maintains that there are spatiotemporal facts or structure not grounded in material bodies. First is what we might call Bainianism: one is a realist about spatiotemporal structure but not about any particular instantiation of it, that is, not about any of the (non-material) objects that could be said to instantiate it. On this view, the different possible descriptions or formulations or instantiations of spatiotemporal structure do not really differ from one another: one is antirealist about those. Second is what we might call uncommitted substantivalism: one is a realist about a particular instantiation of

[32] Rosenstock et al. (2015) suggest that the Einstein algebra formulation is a relationalist formulation of general relativity (which they argue is equivalent in a certain sense to the manifold formulation). By contrast, Earman says that although the Einstein algebra formulation "eschews substantivalism in the form of space-time points," it is "nevertheless substantival, only at a deeper level" (1989, 193). Whether this formulation is more properly seen as relationalist or substantivalist is something I won't explore here. Bain's idea may count as a quotienting view, in the sense of Sider (2020), discussed in the next chapter.

spatiotemporal structure—there is a single best way of describing or formulating the spatiotemporal-structure facts, in terms of a certain kind of non-material object—but one doesn't know what that instantiation or best formulation is, so we can't state the view as propounding one or another such formulation. Third is what we might call committed substantivalism: one is a realist about a particular instantiation of spatiotemporal structure, one thinks that there is a best formulation of it, and one does claim to know what it is; for example, it might be the formulation in terms of points. Fourth is the qualitativist substantivalism of Dasgupta (2009, 2011), on which the fundamental spatiotemporal facts are purely qualitative, not mentioning any entities; spacetime is not an entity but a "purely qualitative structure." The argument below supports substantivalism regardless of which of these versions one ascribes to.

5.6 An argument for substantivalism

I now want to suggest that given this conception of the debate, there is a powerful argument for substantivalism, given much of current physics.

Above I argued that the relationalist (as well as the substantivalist) should adhere to the matching principle by countenancing spatiotemporal structure, and that the relationalist can do this by understanding all the facts about spatiotemporal structure as being grounded in facts about material bodies. (In this way the relationalist can go partway toward adhering to the matching principle. I have not said whether the relationalist can countenance the particular structure required by the laws.)

However, I now want to argue that the relationalist can't really adhere to the matching principle, properly understood.

Recall that the matching principle says to posit in the world the structure presupposed by the fundamental laws, and to posit no more structure than that. It tells us to posit physical structure in the world that matches or corresponds to the mathematical structure needed to formulate the laws.

As with the other epistemic rules governing our inferences about structure, the matching principle applies, in the first instance, to the fundamental laws. (I leave it open whether something similar holds for nonfundamental laws.) Given the *fundamental* laws, we should posit in the world the structure they presuppose. We should posit physical structure in the world corresponding to the mathematical structure needed to support the fundamental laws. This is clear from our usual inferences about spatiotemporal structure. Assuming that Newton's laws are fundamental, we infer that there is a Galilean spatiotemporal structure to the world. Assuming that different laws are fundamental, we ascribe a different spatiotemporal structure to the world—such as a Minkowski spatiotemporal structure for special relativity, a preferred-location spatial structure for Aristotle's physics, or a variety of different spatiotemporal structures for general relativity.

Now here is something else about the matching principle I have not explicitly mentioned. The matching principle tells us to posit, in the *fundamental* level of the physical world, whatever those laws presuppose. The fundamental laws, after all, are about what's fundamental. They don't "care" or "know" about or mention the nonfundamental. I take it this is part of what we mean when we say that they are fundamental. This is a familiar thought. It lies behind our dislike of quantum laws that mention such things as "measurement" or "the observer," for example. (You might think the reason for disliking such laws is that the things they mention are too vague or physically imprecise (Bell, 1990). I agree, but take the imprecision to be a marker of nonfundamentality, and thereby an indicator of something that isn't a candidate for being mentioned by fundamental physical laws.) Fundamental laws will of course have consequences for nonfundamental things; they yield predictions for nonfundamental phenomena when we plug in initial conditions and use various bridge principles. On their own, though, fundamental laws of physics only mention or presuppose or know about things at the fundamental physical level.

Michael Hicks and Jonathan Schaffer (2017) argue against this idea, which they call orthodoxy. They say that fundamental laws can, and often do, mention nonfundamental properties. I agree that such a formulation of the laws can be useful in practice, but I think the best formulation will not mention such things, for reasons in the next section. For now, notice that giving up on orthodoxy would thwart many conclusions we familiarly draw in physics, such as that the second law of thermodynamics, which mentions such nonfundamental things as entropy or heat, is *not* a fundamental law. We typically take the mention of nonfundamental things to indicate either that the law in question is not fundamental, as in the case of the second law of thermodynamics, or that it is a bad formulation of the law, as in the case of quantum mechanical laws that mention "measurement" or "observers" or the like.[33]

Another way to see this comes from the idea of the structure presupposed by the laws. The sense in which the laws presuppose or require some structure is akin to a mathematical idea discussed in Chapter 2. There we saw that higher-level mathematical structures assume or presuppose or constrain lower-level ones, in that higher-level objects or notions cannot be defined until the relevant lower-level ones have been assumed or defined. Higher-level notions don't make sense absent the lower-level ones, and in that way they presuppose them. By contrast, lower-level mathematical structures can be defined independently of higher-level

[33] Arntzenius (2012) suggests that being simply statable in terms of fundamental quantities is a basic starting assumption about the laws: we assume the fundamental laws are (simply) formulable in terms of fundamental things, and this allows us to take these laws as guides to the fundamental nature of the world. This does yield a difficult case I am unsure about. Some have said that the past hypothesis (that the universe began in an extremely low-entropy macrostate) is a fundamental law, even though it is formulated in terms of entropy, which is a nonfundamental quantity. Given the success of the statistical-mechanical underpinnings of thermodynamics, it may be reasonable to expect a natural micro-physical correlate of entropy, but this is a tricky issue: see Callender (1999) for a discussion of the issues involved. See Chen (forthcoming) for argument that fundamental laws can contain vagueness, with the past hypothesis an example.

ones: in this way, the lower levels of structure do not know or presuppose anything about higher-level notions.

In Chapter 3 we saw something analogous for the structure required by the physical laws. Physical laws typically presuppose some structure, in that the structure must be assumed in order for the laws to make sense or be meaningfully formulated. The laws do not similarly know about—require, constrain, presuppose, assume—higher-level structures. For fundamental physical laws, the result is that they only know about fundamental physical structure. (The fundamental laws may constrain things higher up in a different, metaphysical sense: given the fundamental laws and ontology, everything else may be "fixed" in some sense. This is a different sense of constraining from the mathematical one, which concerns what is needed for something to make sense or be defined. The other sense is a metaphysical notion that requires additional metaphysical principles concerning the relationship between different levels of reality.)

An example illustrates and motivates what I will call the primary reading of the matching principle, the one that applies to the fundamental physical laws and the fundamental level of physical reality. Recall the discussion of time reversal invariance in Chapter 3. There I said that if the laws are not time reversal invariant, then we infer that there is an asymmetric temporal structure in the world. For the laws presuppose this structure: they mention or presuppose a distinction between past and future, telling things to behave differently depending on the direction of time. Therefore, given such laws, we should posit this structure.

But there is more to the story, once we explicitly take into account the fact that the principles governing our inferences about structure apply first and foremost to fundamental laws. Thus, take the second law of thermodynamics. This law is not time reversal invariant: it says that entropy can increase to the future, not to the past. Since different things are allowed to happen in each temporal direction, this law seems to indicate an asymmetric temporal structure.

However, the second law of thermodynamics is not a fundamental law. It does not mention systems' fundamental constituents, like particles, and the fundamental features that characterize them, such as their masses and positions. It is formulated in terms of a higher-level macroscopic quantity, entropy. Whether to infer that there is an objective past-future distinction in the world really depends on what fundamental theory accounts for the second law, and whether *that* theory's laws are symmetric in time.[34] The nonfundamental law on its own does not tell us about the world's fundamental temporal structure. It is too far removed from the fundamental physical level to do that.

[34] It is natural to think that if a past hypothesis account of thermodynamics is correct, then there is no asymmetric temporal structure; whereas if a non-time reversal invariant theory such as GRW quantum mechanics is true (and able to account for thermodynamics), then there is. See Albert (2000) on the two accounts of thermodynamics. (Albert does not go on to draw those conclusions about temporal structure.)

In other words: we posit fundamental spatiotemporal structure in the world that's needed for the fundamental laws. We recognize as fundamental the spatiotemporal facts recognized by the fundamental laws. The matching principle applies in the first instance to the fundamental laws and the fundamental level of physical reality. The matching principle as discussed in Chapter 3 says that the structure of the physical world should "look like" or "fit" its fundamental laws. The primary reading of the principle says that the fundamental level of the physical world should look like or fit its fundamental laws. It is this reading of the principle that will spell trouble for the relationalist.

Notice that the kinds of fundamental laws we are familiar with are formulated to presuppose spatiotemporal facts apart from material bodies. They mention or presuppose a spatiotemporal structure in addition to material bodies and their features. The laws of Aristotle's physics mention a preferred-location spatial structure as well as the elements that move toward their natural places. Newton's laws, standardly formulated, presuppose a Galilean spatiotemporal structure and also refer to massive bodies. The laws of special relativity assume a Minkowskian spatiotemporal structure in addition to particles and fields. The field equations of general relativity, on the usual understanding, say how the distribution of matter and energy relates to the spatiotemporal geometry, which in turn affects the behavior of matter. These equations are formulated directly in terms of, they directly mention or talk about, a spatiotemporal structure apart from material bodies, coded up in the metric tensor. (See Hoefer (1996, 1998) for argument that the metric is most naturally seen as characterizing a spatiotemporal structure that is not the structure of a material field. This is not uncontroversial, but it is assumed in standard presentations.)

The fundamental laws we are most familiar with are formulated to directly mention material bodies, with terms that directly refer to them, such as the mass term of Newton's dynamics, or the mass density of some formulations of Newtonian gravitation, or the elements mentioned in Aristotle's laws, or the stress-energy tensor of general relativity. At the same time, these laws are formulated in such a way as to presuppose or make reference to a spatiotemporal structure apart from those bodies—apart in that this is presupposed by the laws in the mathematical sense noted above, or else is directly mentioned by or coded up in a distinct term.[35] (I turn to potential reformulations of the laws in Section 5.7.)

This yields a difficulty for the relationalist. The problem isn't that of failing to recognize enough spatiotemporal facts for the physics, a concern lying at the root of classic arguments like Newton's as well as many contemporary ones. Grant the relationalist enough stuff to ground those facts and to make the relevant predictions, and there is still a problem. According to the core of the view, all

[35] In the context of this debate, both views take certain material objects to exist at the fundamental level. Supersubstantivalism would then deny this.

the facts about spatiotemporal structure are grounded in more fundamental facts about material bodies. Yet the kinds of fundamental laws we are used to presuppose or mention spatiotemporal facts apart from material bodies—facts that, for the relationalist, are nonfundamental. This violates the principle that the fundamental level of the physical world should contain whatever is needed for or presupposed by the fundamental physical laws.

So the argument is this. First premise: the fundamental laws are about what's fundamental to the physical world; they refer to or presuppose things about the fundamental physical level. Second premise: these laws are about, they presuppose or refer to, a spatiotemporal structure or spatiotemporal facts apart from material bodies. Third premise: for the relationalist, this kind of structure or fact exists at a nonfundamental level, above that of material bodies. Fourth premise: the primary reading of the matching principle. Conclusion: relationalism is incorrect.

General relativity provides an illustration. The theory establishes a nomological connection between material things such as matter and fields, and a spatiotemporal structure apart from them. On their own, these laws do not say whether material things and spatiotemporal structure are at the same level of physical reality, nor which is more fundamental if not. Without some further principle, both relationalism and substantivalism seem satisfactory: both views can countenance facts about material bodies as well as a world's spatiotemporal structure. Enter the matching principle. The substantivalist, but not the relationalist, adheres to it.

I think this captures the gist of what the substantivalist has been thinking all along: that the theories we are most familiar with are about spacetime as much as material bodies. In my terms: the laws presuppose a fundamental spatiotemporal structure not grounded in material bodies. That said, as I discuss below, a future physics, with very different laws, could suggest otherwise.

A worry. Suppose that what I have been calling spatiotemporal structure involves, at least in part, facts that must be stated using universal generalizations. On a standard axiomatic approach to geometry, for example, a given spatiotemporal structure will be defined by means of a universal generalization over a domain of points. Suppose further that generalizations are not fundamental but are grounded in their instances, in accord with a familiar way of thinking about grounding. Then it may seem as though the substantivalist does not adhere to the matching principle either, simply because spatiotemporal structure, qua generalizations, cannot be fundamental.

The substantivalist will avoid this worry, for one of the following reasons. First, one might for independent reasons think that generalizations are fundamental, a not-unprecedented (to my mind, not implausible) view, even among grounding proponents. Second, even if spatiotemporal structure-qua-generalizations is not absolutely fundamental, it is still *very close* to being fundamental, so that the fundamental structure of the world *almost directly* matches the structure for the fundamental laws. The only "gap" there is between spatiotemporal structure and the fundamental level is the one created by the gap between generalizations and their instances. This is an intuitively smaller gap than that between a world's spatiotemporal structure and the features of material bodies. The former is just a "gap in logical form"—the "size" of the separation between a generalization and the collection of particular claims that grounds it; the latter is a larger, physical gap. The substantivalist thus adheres to the matching principle more than the relationalist does. Finally, even if the generalizations that axiomatize a given structure are not absolutely fundamental, the various facts about the points still can be, and these facts are included in my conception of spatiotemporal structure; in which case there are still fundamental spatiotemporal facts or structure apart from material bodies.

(The worry would also seem to go too far. It would force us to say that no particular collection of fundamental facts is to be preferred to any other on the basis of the physical laws, simply because any structure required for the laws takes the form of a generalization, and no generalization is fundamental. But surely a matching-type argument can sometimes work, as when we want to say that Berkeleyan idealism posits a world that radically fails to match the structure indicated by the laws. Surely we may reject that view for this reason—even though the structure indicated by the laws is given by generalizations, and even if generalizations are not fundamental but grounded in their instances.)

Note that the argument for substantivalism is independent of one's view on the metaphysics of laws. The question of what makes a statement a law is distinct from the injunction to posit, assuming that a certain statement is a law, the requisite structure in the world. Even the Humean, who denies that laws of nature are metaphysically fundamental, can agree to posit, in the fundamental physical level of the world, the structure presupposed by the fundamental physical laws. To put it another way, the content of the law claim, the proposition *p* of the statement "it is a law that *p*," is what indicates structure in the world. It is irrelevant whether what makes it the case that *p* is a law is itself metaphysically fundamental. Whatever your account of laws of nature, that is, you can—and should—adhere to the matching principle.

The matching principle is a familiar and successful guiding principle. It applies in the first instance to the fundamental laws and the fundamental level of physical reality. The substantivalist and relationalist, as I see them, disagree about the nature of the fundamental physical level, which is why the matching principle can distinguish between them. This is a substantive debate about the fundamental nature of the world according to physics; a debate about what makes it the case that the spatiotemporal structure required by the physics holds. It is a debate that, given current physics, the substantivalist seems to be winning.

That said, it remains open for future physics to turn the tide. If a quantum theory of gravity or some other future fundamental theory contains laws that only presuppose things about material bodies and their relations, which in turn give rise to the spatiotemporal structures presupposed by our current theories, we may conclude that relationalism is correct. On a natural conception of certain approaches to quantum gravity, spatiotemporal structure does not seem to be fundamental, but in some sense emerges from something else fundamentally non-spatiotemporal. The approach known as causal set theory, for one, posits at the fundamental level a discrete set of events that are partially ordered by a causality relation. There is only a causal structure, or causal facts, at the fundamental level, which is said to give rise to the spatiotemporal structures or facts assumed by current physics. This may be a relationalist theory, on my understanding. As Nick Huggett and Christian Wüthrich (forthcoming, Ch. 2) put it, the entire goal of the approach is "to take causal structure as fundamental and show how this structure grounds everything else about spacetime." (Throughout their book, Huggett and Wüthrich argue that various programs in quantum gravity suggest that spatiotemporal structure is nonfundamental.[36]) Future laws might even suggest a view that does not look like either relationalism or substantivalism, presupposing facts about neither material bodies nor spatiotemporal structure, but something else altogether. Wüthrich (2019) concludes something like this when he notes that, of the theories of quantum gravity he surveys (causal set theory and loop quantum gravity), spatiotemporal structure does not appear to be fundamental—but that neither do material bodies and their relations.

Hoefer says that, "substantivalism and relationism today must be understood in part as bets" (1998, 462) about how future physics will turn out. The argument in this section suggests that substantivalism is on track to win that bet, while at the same time allowing that future physics might indicate otherwise. (You might think the previous paragraph belies the suggestion that substantivalism is on track to win the bet. But given how speculative the various programs in quantum gravity are at this stage, it remains safe to say that one should bet on substantivalism on the basis of our most familiar and well-understood theories. And the arguments of Huggett and Wüthrich notwithstanding, there is room to question whether causal set theory and other approaches do postulate a wholly non-spatiotemporal structure at the fundamental level; cf. Esfeld (2019). An investigation into different programs in quantum gravity is beyond the scope of this book.) This conception of the debate is not purely metaphysical, merely verbal, or otherwise divorced from physics, in other words. It is relevant to current physics, and will continue to be relevant to future developments in physics.

5.7 A challenge for relationalism

The argument above assumes that various candidate fundamental laws we currently have, for physical theories from Aristotle's physics to general relativity and

[36] The dominant viewpoint nowadays seems to be that spatiotemporal structure is emergent in quantum gravity, though it is not universal. Esfeld (2019) is one dissenting view.

more between, are formulated to presuppose a spatiotemporal structure apart from material bodies. This reveals a different way for the tide to turn: the relationalist could reformulate these laws so that they only presuppose things about material bodies. If such a reformulation is possible, then the argument will turn on how we should generally formulate the laws, which is a big question I cannot fully answer here. That said, the argument poses a significant challenge to any relationalist attempt to reformulate the laws.

To see why, consider an illustrative example: the relationalist reformulation of Newtonian mechanics suggested by Bas van Fraassen (1970, Sec. 4.1) and filled out in one way by Nick Huggett (2006). Van Fraassen and Huggett say that we can reformulate Newtonian mechanics to include the statement that "Newton's Laws hold in some frames," where these will be the inertial frames. (There will also be a force law, and on Huggett's version of the theory, a law about the spatial geometry.) These laws pick out a standard of inertia without assuming that spacetime exists. In my terms, they only presuppose spatiotemporal facts about material bodies. This is because, according to Huggett, the facts about inertial frames, indeed all the spatiotemporal facts, themselves supervene on facts about the history of relations between material bodies.

This is a genuinely relationalist formulation, as I see it, which respects the primary reading of the matching principle. The truth of the laws in certain frames effectively substitutes for an inertial structure, so that the laws themselves do not have to mention or presuppose this structure. In Huggett's words, "it is not the structure of absolute space that makes certain frames privileged, just the truth of the laws in those frames" (2006, 46). (Notice that Huggett must have in mind a reasonably fine-grained conception of the laws. As mentioned in earlier chapters, Newton's laws can be reformulated so as to apply to non-inertial frames, with equations that contain additional pseudo force terms. In order for the truth of Newton's laws to pick out the inertial frames, Huggett must hold that the reformulated equations do not count as laws; otherwise, the truth of the (reformulated) laws, in those frames, would pick out the non-inertial frames as well. He must be assuming that there is a means of distinguishing between the original and reformulated laws, on the basis of which he can say that the reformulated equation is not, in fact, a genuine law. Incidentally, this goes against one thing often said in physics books, which is that the reformulated equations are just different expressions of the laws: they are still laws, indeed the same laws, just stated differently.)

The problem is that this seems like a worse formulation of the laws, for a few reasons. First, Huggett's formulation does not respect the idea that fundamental laws only mention fundamental things. For the laws are given in terms of facts about inertial frames, which for Huggett are not fundamental but grounded in facts about the relations between material bodies. The thought that such a formulation of fundamental laws is worse is not uncontroversial (recall that Hicks and Schaffer (2017) argue against it), but it is reasonably standard, and reasonably compelling.

Arntzenius (2012) goes so far as to take it as a starting point of our theorizing that the fundamental laws are simply formulable in terms of fundamental properties and relations, since without this assumption "one gets nowhere... when trying to extract the fundamental structure of the world from theories of physics," as it says on the back cover of his book. By the same token, laws that are simple when formulated in terms of things the theory regards as fundamental is evidence of a good theory. Dorr notes that,

> When we are trying to figure out how to divide our credence in a reasonable way between hypotheses about fundamental structure, considerations of simplicity will matter a lot. What we want is not just a short list of fundamental properties and relations, but a simple set of laws stated in terms of these properties and relations. (Dorr, 2011, 146)

Huggett's reformulation of Newton's laws lacks this epistemic virtue. For that reason, it is less preferable to the original. (Dorr's and Arntzenius' focus is on the simplicity of the laws when stated in terms of fundamental things. I am emphasizing that the laws are not stated in terms of things the theory regards as fundamental at all, as required to be able to apply their measure of simplicity.)

Huggett acknowledges that his formulation is not given solely in terms of fundamental things (or natural properties, as he puts it), but it seems to me this should strike him as more problematic than he allows. He says that the "real challenge" for any relationalist reformulation is to avoid simply positing the requisite quantities as primitive, since this makes the formulation seem like "a blatant cheat" (2006, 46). (A charge he levels at Lawrence Sklar's idea of a primitive monadic property of acceleration in particular: note 8.) It is in order to avoid the cheating objection that Huggett says the facts about inertial frames supervene on the history of relations between material bodies: the inertial facts are not unexplained primitives, but hold in virtue of relationalist facts.

This still feels like cheating in a relevant sense. Since the laws are not formulated in terms of things the theory takes to be fundamental, it seems like the theory isn't playing straight with us. One wants to see evidence that relationalism is an accurate portrayal of the fundamental nature of the physical world, and one source of evidence for such a thing comes from laws formulated in terms of fundamental relationalist quantities. Otherwise, one suspects that it may not be possible to formulate the laws in a simple way in terms of the alleged fundamental things; and if that were the case—if the only way to formulate the laws in terms of relational quantities is enormously complex and hence insufficiently law-like—why should we believe that the quantities claimed to be fundamental really are so? (This is not to say that the laws being simply statable in terms of a property is sufficient to indicate that the property is fundamental, but it is plausibly defeasible evidence of it.) One should not simply stipulate that the requisite facts are primitive without further ado, but neither (other things being equal) should one formulate the laws in terms of nonfundamental things. That, as Huggett himself says of the "blatant cheat," feels like theft over honest toil. The worry in the background is that it will be too easy to come up with laws based on any fundamental nature of the world one likes, if one is allowed to help oneself to any kind of quantities in formulating them. (Recall the discussion of Earman's "cheap instrumentalist rip-off" from Chapter 1.)[37]

Another reason Huggett's formulation seems worse is that it is given in terms of reference frames, in particular the existence of a certain type of reference frame, explicitly mentioning the existence of such frames. The concern here is that fundamental physical laws are best formulated in terms of things that are directly about the physical world, and reference frames don't fit the bill. According to Newton's laws, inertial reference frames are like units of measure or coordinate systems in that the choice of inertial frame is an arbitrary choice in description. Now, Huggett's formulation doesn't mention any particular frame, nor does it directly mention the inertial ones. Instead it says that there are frames you can choose such that Newton's laws are true, and these will turn out to be the inertial ones. (Recall that assuming a preferred frame for Newton's laws amounts to assuming an absolute space structure, which is bad because this assumes structure in the world that isn't needed for the laws. This is not what's going on here, since no particular frame is preferred.) Nonetheless, the fact that a choice of inertial frame is arbitrary suggests that inertial frames in particular, and reference frames in general—these objects as a group or kind of thing—are descriptive or labeling devices, not inherent in physical systems themselves. Hence they should not, other things being equal, be mentioned or referred to in the fundamental physical laws.

An idea from Field helps bolster the thought that such a formulation is worse. Field draws a distinction between what he calls intrinsic and extrinsic explanations.[38] Intrinsic explanations "explain what is going on without appeal to extraneous" entities, things that are "*extrinsic to the process to be explained*" (1980, 43; original italics). The result is that intrinsic explanations are better, more "illuminating" (1980, 43) or "satisfying" (1989, 18):

> [E]xtrinsic explanations are often quite useful. But it seems to me that whenever one has an extrinsic explanation, one wants an intrinsic explanation that underlies it: one wants to be able to explain the behaviour of the physical system *in terms of the intrinsic features of that system*, without invoking extrinsic entities . . . whose properties are irrelevant to the behaviour of the system being explained. If one cannot do this, then it seems rather like magic that the extrinsic explanation works. (Field, 1989, 193; original italics)

[37] Compare Arntzenius (2012, 169–70).

[38] Field (1989, 193 n. 33) cites Loar (1981, 62) for this terminology.

The best explanations are intrinsic, citing features that are directly relevant to a system's behavior. Extrinsic explanations, though useful, seem "like magic."

Draw an analogy to different types of formulations of the laws. Formulations of the laws in terms of reference frames or coordinate systems or other such descriptive devices are similarly extrinsic: they mention things outside physical systems themselves, referring to the labeling devices we impose upon them. Such formulations are then worse for the same reasons Field says that extrinsic explanations are worse. They reference things whose properties aren't directly relevant to physical systems and the behaviors we want to account for. (Compare Einstein on a coordinate system, which is "only a *means of description* and in itself has nothing to do with the *objects to be described*" (2002, 203; original italics): coordinate systems *in themselves* do not concern physical systems, as evidenced by the arbitrariness they involve, even though they can be used to successfully describe physical systems.[39]) Extrinsic formulations of the laws can be useful, yet one also "wants an intrinsic formulation that underlies it," which will help explain the success of the extrinsic one. If it is not possible to do this, then the success of the formulation seems like magic.

Other things being equal, we should prefer an intrinsic formulation of the laws—or what I have been calling in this book a *direct* formulation. Things like coordinate numbers and reference frames can tell us about physical systems, so that the formulation needn't seem like complete magic. It is just that they do so in a less direct, and therefore less preferable, way.

Recall a similar idea for mathematical objects. We can characterize the structure of the Euclidean plane by saying that there are coordinate systems in which the distance formula takes the usual Pythagorean form. Notwithstanding the mention of coordinates, it is not magic how this description works, since a plane allows for such coordinate systems just in case it has that structure. Still, the reason for the success of the characterization in terms of coordinates flows from the plane's structure. It *would* seem like magic if the characterization bottomed out at the existence of certain kinds of coordinate systems, since coordinate systems are labeling devices not inherent to the space itself. The existence of a certain type of coordinate system cannot be a bottom-level fact about the space, but must hold in virtue of its intrinsic nature or structure.[40] The better—more direct, perspicuous, explanatory, fundamental—characterization of the plane's structure is given by

[39] Einstein's own thinking about coordinate systems was more subtle than this one quotation suggests, and it changed over the years; see Norton (1993a). Immediately after the above quotation, Einstein goes on to say that, "Only a law of nature in a generally covariant form can do complete justice in this situation, because in any other way of describing, statements about the means of description are jumbled with statements about the object to be described," which is a further conclusion we have seen reasons to be wary of.

[40] In drawing this asymmetry between coordinate-based and coordinate-independent descriptions of a given structure, suggesting that one is more fundamental than the other, I depart from Wallace (2019), whose defense of coordinate-based descriptions I otherwise agree with.

a geometric object, a metric tensor,[41] which does not mention coordinate labels and more directly captures the nature of the plane, itself a geometric object. The characterization in terms of coordinate systems does specify the structure of the plane, but in a way that is needlessly indirect—"intolerably roundabout," to borrow a phrase from Newton (see the epigraph to this chapter)—in terms of the kinds of coordinate systems we can lay down on top of it.

Likewise for the formulation of a fundamental physical theory in terms of reference frames. Such a formulation can tell us about the world, and the mention of reference frames in Huggett's hands does not make the formulation frame-dependent in the sense of requiring a particular choice of frame or changing in truth value with a change in frame. Nonetheless, the formulation is "intolerably roundabout." It imposes an additional "screen of mathematical representation between us and the object in which we are interested," in Maudlin's phrase (2014a, 9), potentially obfuscating the nature of the object in question. (An additional screen: there is bound to be some such screen, as a physical theory is not itself constructed out of physical entities.) Better to have a formulation in terms of things that are more directly about the physical world.

This is especially true when we are in the business of trying to figure out the nature of the world according to a physical theory. Otherwise, we can be misled into thinking that the theory is really about those things—that, say, "More than anything else, the special theory of relativity is a theory about reference frames" (Susskind and Friedman, 2017, 3)—just as we can be misled into thinking that quantum mechanics is really only about "measurement" or what's "observable," given the formulation of the laws one finds in standard textbooks. As Bell complains of such formulations of quantum mechanics: "It would seem that the theory is exclusively concerned about 'results of measurement', and has nothing to say about anything else" (1990, 19). All other things being equal, it is better to have a formulation that will not mislead us into thinking that the theory is only or centrally about things like measurement or reference frames. In other words: all things equal, it is better to have a direct formulation, which will be more metaphysically perspicuous, and so less likely to mislead us about the true nature of the physical world. (It is not uncommon for physics books to state the laws in terms of reference frames or coordinate systems. The claim is that this is not in general the best formulation.)

Let me reiterate that we can learn about the nature of the physical world from a theoretical formulation that mentions reference frames, just as we can learn about the nature of the Euclidean plane from a characterization that mentions coordinates. Nonetheless, when choosing among different theories or formulations for

[41] Or a family of metric tensors, each of which captures the structure of the space up to choice of unit: recall from Section 2.4.

the purposes of gleaning the fundamental nature of the physical world, we should, other things being equal, prefer the one that is more direct.

This does not go against the tenor of Chapter 4. The formulation of Newton's law discussed there is direct, given the standard metaphysics of the theory. The formulation of the Lagrangian equations is likewise relatively direct. As mentioned in Chapter 4, generalized coordinates are functioning as placeholders in the equations, allowing us to state the laws without having to directly refer to any particular coordinates. By contrast, Huggett's formulation of Newton's laws explicitly and essentially and directly refers to the existence of certain reference frames. In effect, the equations of Lagrangian mechanics use generalized coordinates but do not mention them; Huggett's laws explicitly mention, they directly refer to, reference frames.

In saying that direct formulations of the laws are preferable, I do not mean to suggest that we should aim to nominalize physics (or that a characterization of the plane in terms of Euclid's axioms is preferable to one in terms of a metric tensor). One could take the idea this far, but I wish to leave that aside. I only mean to distinguish between formulations given in terms of coordinate systems or reference frames and the like, and formulations not given in terms of such things—to distinguish between more and less direct formulations. This is a distinction between different types of abstract mathematical formulation. I am not concerned to distinguish between formulations of the laws that mention numbers or other abstract entities and those that do not.

You may continue to wonder why we should think that a direct, metaphysically perspicuous formulation of the laws is better. It is hard to say why exactly, but the underlying thought is akin to the cheating objection. Think of the instrumentalist who says that there aren't really any sub-atomic particles, even though things behave as if there were. A natural reaction is to feel that this cries out for explanation. Why do things conspire to behave exactly as if there were sub-atomic particles? The best answer is that there really are sub-atomic particles, which behave in the ways the laws claim. It is cheating to say that everything behaves as if there were such things, and to leave it at that. Similarly here. Why do the laws hold in certain reference frames? An inadequate answer is to say that the laws simply prefer those frames; that the laws conspire to make it seem as though there is something in the world underlying the preference for certain frames. That answer feels unsatisfying, contrived.[42] The best answer directly refers to the nature of the world that makes those the frames in which the laws hold.

Direct formulations may be preferable only if you are a realist, or at least an anti-instrumentalist, to begin with—only if you think it is the job of a physical theory

[42] One way to flesh this out is to argue that the laws construed in such a way would not be explanatorily satisfactory: Dorr (2010, 2011, Sec. 7). Recall the discussion of the explanatory superiority of direct formulations from Chapter 1.

to tell us about the nature of the world. The instrumentalist may be unbothered by indirect formulations and extrinsic explanations. (The instrumentalist should be used to the charge that the success of science seems like magic.) Since it is not my aim to argue against instrumentalism here, I will leave it to my opponent to parry the objection that such formulations are worse. I will note, though, that indirect formulations seem particularly problematic for fundamental physical laws, since the elements that feature in them (reference frames, coordinate systems, or the like) do not seem like the sorts of things that can be truly fundamental or explanatory.

There are other relationalist reformulations to consider in more detail than I can do justice to here. However, the above strikes me as indicative of the kinds of problems that any such reformulation will face. In order for relationalism to be victorious, the proffered reformulation must be genuinely relationalist, presupposing facts only about material bodies; it should be direct; and it should respect the primary reading of the matching principle—all the while containing laws that have the hallmarks of genuine laws of nature (simplicity and generality and the like). It is hard to think of a relationalist reformulation of our current laws that will satisfy all these criteria. (Though here is one potential example. The Newtonian two-body problem, within the framework of Lagrangian or Hamiltonian mechanics, can be formulated using relative distances as the only configuration variable (while effectively encoding angular momentum in a potential energy term).[43] Such a formulation is relationalist, as I see it, and it appears to meet the above criteria. The remaining question is whether it can be sufficiently generalized beyond the two-body problem, something worth investigating; cf. note 46.)

A brief look at a few more examples further suggests that a relationalist reformulation meeting these criteria will be hard to come by. (This is by no means an exhaustive list, nor will I examine these in exhaustive detail.)

(1) Julian Barbour's relationalist formulation of mechanics (Barbour and Bertotti, 1982; Barbour, 1982, 1999) aims to do away with any fundamental temporal or spatial structure. The result is an indirect formulation of mechanics. The indirectness enters in recovering the topological temporal structure and the inertial spatiotemporal structure, as the discussion in Arntzenius (2012, Ch. 1 and Sec. 5.11) makes clear. For example, Barbour aims to do away with any fundamental temporal structure, even a topological structure, by reformulating the laws as, in Arntzenius' explication, "there exists a way of ordering the instants so that certain dynamical laws (which presuppose such an ordering) are true relative to that ordering" (2012, 32). There is no fundamental temporal structure (not at the level that would count as a temporal structure), yet systems' physical states can be ordered as if there were; and relative to that ordering, the given laws will hold.

[43] Thanks to Gordon Belot and Laura Ruetsche for the example.

These laws pick out the relevant ordering without presupposing a fundamental temporal structure: the formulation is genuinely relationalist about temporal structure. Yet by the same token, the ordering referred to in the laws is not inherent in the histories of physical systems. It not directly about physical systems and their behavior, but is something we impose on them, a labeling device used to keep track of a particular ordering of states that is not intrinsic to systems' histories. We can see this in the words of Barbour and Bruno Bertotti:

> This is Leibniz's concept of time as merely the successive order of things: instants are defined by the successive relative configurations of the Universe.... These define a curve in Q_0 [the relative configuration space, specified via particles' relative positions] whose points can be labelled by a monotonic and continuous parameter λ, a purely topological label. (Barbour and Bertotti, 1982, 296)

They conclude that such a formulation "dispenses with an independent time" (1982, 296). The formulation is relationalist, but it is also indirect. That is a reason to prefer the original.[44]

(2) Huggett mentions another law of his reformulation of Newtonian mechanics: "'there is an embedding of the relational history into G', for some specific Riemannian geometry G," so that, "to say that a distribution of matter is (geometrically) 'possible' is just to say that it is embeddable in this space" (2006, 53). For Huggett, the privileged embedding supervenes on the history of relations between material bodies, just as the inertial frames do. Facts about the embedding geometry—about the spatial structure—are not fundamental but are grounded in facts about material bodies.

This law, too, explicitly mentions things the theory regards as nonfundamental: a spatial structure that holds in virtue of the behaviors of material bodies. A similar charge applies to any reformulation on which the laws say something like, "A given history of changes in the distances between certain particles is physically possible if, and only if, it can be conceived to take place within Newtonian absolute space in such a way as to satisfy $F = ma$" (Albert, 2018). Such laws state that particles behave as if they existed in a space with a certain structure, thereby mentioning a spatial structure without presupposing it. Such a formulation is not substantivalist, as I see it, but at the same time it mentions structures that are not regarded as fundamental.

[44] A similar worry arises for Barbour's "best matching principle" that is used to do away with a fundamental inertial structure: see Arntzenius (2012, Ch. 1, esp. p. 32 n. 16). The theory further seems to presuppose a spatial structure over and above that of material bodies, as suggested by the presentation in Earman (1989, Secs. 2.1, 5.2) (Arntzenius (2012, Sec. 5.11) and Pooley (2013, Sec. 6.2) suggest this of Barbour's reformulation of general relativity in particular), in which case it counts as substantivalist about spatial structure, on my understanding.

Albert suggests that it is not quite right to construe such a formulation this way.[45] He allows that the "as if" locution seems like cheating, as though the relationalist law invokes what is merely a fictional entity. Instead, he says that the structure is simply being used as a concise way of expressing the constraints on inter-particle distances. Fair enough, but this raises a relevantly similar concern. If the only way to concisely specify the constraints on inter-particle distances is by reference to a certain spatial structure, then that in itself is reason to believe that the structure is fundamental. For it is presupposed by the best (simplest, most concise, and so on) formulation of the laws. It still feels like cheating if it cannot be shown that the laws are formulable concisely in terms of what the theory takes to be fundamental.

(The dynamical approach of Brown and Pooley (Brown (2005), Brown and Pooley (2006)), on which the facts about spatiotemporal structure are less fundamental for purposes of explanation than the symmetries of the laws, might fall into this category, depending on further details. Put in that way, the view is orthogonal to the relational-substantival debate, as Brown himself says (though Pooley says that it "qualifies as a type of relationalism" (2013, 569)). At times Brown and Pooley suggest that the laws and their symmetries hold in virtue of the behaviors of material bodies, in which case the view seems to be relationalist. Huggett and Hoefer (2009) say the view entails the problematic "as if" type of claim: that material bodies behave as if they are embedded in a background spacetime.)

(3) Another example is the account in Albert (2019b, 2020). Albert reformulates the laws so that they do not presuppose or require a spatiotemporal structure. (Recall that in his view there is no such thing as a fundamental pre-dynamical spatiotemporal structure.) On this formulation, the laws say that there exist certain coordinate systems in which the laws take a particular form and relative to which certain distributions of material bodies are allowed. To say that the laws are Maxwellian, for example, is just to say that there is some coordinatization of the manifold relative to which the laws take a Maxwellian form; and relative to that coordinatization, certain distributions of material particles and fields are solutions. (The form of the laws relative to those coordinates is then what gives rise to a particular spatiotemporal structure.) These laws only presuppose a topological and differentiable structure (which Albert allows to be fundamental: note 17), and they do not invoke a fictional spacetime or entail any problematic "as if" claim. This formulation is genuinely relationalist, and it respects the primary reading of the matching principle.

The reference to coordinate systems nonetheless makes the formulation indirect. It is akin to specifying the structure of the Euclidean plane by reference to the existence of coordinate systems in which the metric takes a particular form. This

[45] In a seminar at Rutgers in Spring 2017.

suffices to characterize the plane, but in a way that feels "intolerably roundabout." One wants a direct formulation of the laws, and a direct characterization of the plane, which will explain why there exist such coordinate systems. Other things being equal, it is preferable to have a more direct formulation. (Again, a concern in the background is that this type of maneuver can be used to eliminate any kind of thing one wants, even with laws that have a superficially simple form, as argued by Dorr (2010, Sec. 6), who suggests that in general, "Weakening a theory by 'existentially quantifying out' some putatively structure-characterizing predicates makes it worse" (2010, 166).)

This does not prove that no relationalist reformulation can succeed, and more work must be done to fully evaluate the various proposals on offer in these terms, including the assorted programs in quantum gravity, relationalist construals of general relativity, and various relationalist reformulations of classical mechanics.[46] However, the above suggests that it won't be easy to find a relationalist reformulation that has the features we want of fundamental laws. The sorts of candidate fundamental laws we are most familiar with are formulated to presuppose a spatiotemporal structure apart from material bodies. The problem is that the typical relationalist substitutes for that structure, facts about things like reference frames or coordinate systems or embedding geometries, are not the kinds of things that appear in direct formulations of the laws stated in terms of fundamental relationalist quantities. That said, the laws of a future physics could turn out differently.

In any case, by reframing the traditional debate in this way, we manage to locate a new framework that opens things up to fruitful investigation from both physics and philosophy. So that even if the traditional dispute has stagnated or become non-substantive, there is still an interesting and substantive question here. It is not exactly the question debated by the likes of Leibniz and Newton, but it is close in spirit. It is a debate about the fundamental nature of the physical world, something that is clearly relevant to, and will continue to be informed by, current and future physics.

[46] Some examples of proposals worth further investigation: the relationalist statespace formulation of classical mechanics in Belot (1999, 2000); the Einstein algebra formulation of general relativity (note 32; see Arntzenius (2012, Sec. 5.11) for reasons to think it will not be an improved theory, even if it does count as relationalist); the continuum mechanics of Wilson (1993) (which has spatiotemporal structures "attached to the matter" (1993, 217)); the Barbour-style reformulation of Bohmian mechanics of Dürr et al. (2020); arguments that according to quantum mechanics, ordinary spacetime emerges from a very high-dimensional space (Ney and Albert (2013); Albert (2019a); Ismael (2020)).

6
Realism about Structure

> The aim of Lagrange was, as he tells us himself, to bring dynamics under the power of the calculus, and therefore he had to express dynamical relations in terms of the corresponding relations of numerical quantities.... We must therefore avail ourselves of the labours of the mathematician, and selecting from his symbols those which correspond to conceivable physical quantities, we must retranslate them into the language of dynamics. In this way our words will call up the mental image, not of certain operations of the calculus, but of certain characteristics of the motion of bodies.
>
> James Clerk Maxwell (1890d, 308)

6.1 Introduction

I have been emphasizing a certain realism about structure, both mathematical structure in a theory's formalism and physical structure in the world. We should take the mathematical structure required by our best physical theories seriously in telling us about the physical world, as part of a general realism about our best scientific theories.

This emphasis on structure may seem to have some untoward consequences, including a naive method for interpreting scientific theories and an overly strict criterion for the identification of theories and the (non)equivalence of different theories. Rather than agreeing with my conclusions about classical mechanics, for instance, you may want to reject the structural considerations they are based on.

In this chapter I argue that the view has no untoward consequences. The truly radical apparent consequences are not in fact consequences of the view, and the genuine consequences are reasonable things for the realist to hold.

6.2 Taking the mathematics (too) seriously

I have suggested that we regard the mathematical structure required to support the fundamental laws as representing genuine physical structure in the world. The relevant mathematical structure may represent physical structure more or less directly (more, in the case of spacetime structure; less, in the case of statespace

Physics, Structure, and Reality. Jill North, Oxford University Press (2021). © Jill North.
DOI: 10.1093/oso/9780192894106.003.0006

structure). Regardless, we should take this structure seriously as part of a theory's genuine representational content.

This may seem like a simpleminded attitude toward the mathematics in which our physical theories are formulated. It may seem like a "fetishism of mathematics," as Stachel calls it, "the tendency to assume that all the mathematical elements introduced in the formalization of a physical theory must necessarily correspond to something meaningful in the physical theory and, even more, in the world that the physical theory purports to help us understand" (1993, 149). It may seem that I take the mathematics in which our theories are couched altogether *too* seriously.

Although I do suggest that we take the mathematics seriously, nothing in what I have said implies that we must take every mathematical feature of a formulation to (directly or indirectly) represent physical features of the world. There can be mathematical aspects that are "mere gauge," mere artifacts of description that do not correspond to anything physical. In Lagrangian mechanics, for example, we may use a particular coordinate system for the purposes of describing a system, but this does not mean that all the features specific to that choice, such as where the origin is located, correspond to genuine physical features: given the physics, we know that where we place the origin is an arbitrary choice made on the basis of descriptive convenience. ("Given the physics": with a different physics, such as Aristotle's, this may not be completely arbitrary in the sense that there is a particularly natural choice.) Similarly, the laws of Lagrangian mechanics do not say different things about the world when stated in terms of one set of generalized coordinates as opposed to another. (A paradigmatic example of a descriptive choice usually taken to be mere gauge concerns the potentials in classical electromagnetism, discussed in Chapter 7.) At the same time, the structure required to formulate the Lagrangian laws in terms of generalized coordinates, in terms that abstract away from the particulars of any given choice of coordinate system, does tell us about the world according to the theory.

Indeed, my conception of structure assumes that there are such artifacts of mathematical representation. A theory's structure consists in those features that the different allowable representations all agree on, the ones in virtue of which we may reasonably conclude that the different representations *are* all equally legitimate. To say that different mathematical descriptions all represent physical reality equally well is just to say that there are differences in mathematical representation that do not correspond to genuine physical differences. It is to say that there are mere descriptive artifacts in our mathematical representations of physics.

It may strike you that I have not offered a precise or rigorous account of exactly which features of a formalism do, and which do not, correspond to genuine features of the physical world. In Chapter 2, I discussed some general ways of distinguishing between mathematical features that represent genuine structure and ones that do not. I suspect there is no more precise account to be had, since it will depend to some extent on the vagaries of interpretation, in ways I address in the next chapter. But we do not need a definitive rule for when a mathematical feature corresponds to a physical one in order for the basic point to stand, which is that there are good reasons to take certain types of mathematical features, and not others, to correspond to genuine physical features of the world; *ipso facto* there are some mathematical features that do, and others that do not, correspond to genuine physical features of the world.

You might still feel that this amounts to a naive reification of the mathematics. Even if I elude the charge of taking all the mathematical features of a formulation too seriously, it remains the case that I do take certain features very seriously indeed, and it may seem as though I adopt a simpleminded realism with respect to those. For those things, it may even seem that I go so far as to confuse the mathematical representations with the physical things being represented, when of course mathematical objects are very different sorts of things from physical ones, with very different kinds of features. We must remember to mind the gap between the mathematics and the physics, to "keep in mind the distinction between the representational vehicle (the mathematics) and the thing in the world it represents," as Alisa Bokulich (2020, 185) puts it. Maudlin likewise reminds us to "keep the distinction between mathematical and physical entities sharp" (2013, 129). Otherwise, we can be easily misled into thinking that all the features of a mathematical representation directly correspond to physical reality—into thinking that there is a physically preferred spatial location solely because we happen to have arbitrarily chosen a coordinate system with a particular origin, say. It may seem as though I am running roughshod over this basic distinction.

My realism about structure does not deny any of this. Nothing about the view impels us to blithely ignore the distinction between the mathematical structure in which a theory is formulated and physical structure in the world. The view simply says to take the former as an epistemic guide to the latter—thereby presupposing that there is a distinction between them. This does mean that we must be careful to pay attention to which mathematical features are genuinely representational and which are not. But that is an admonition to be careful, not a reason to avoid taking the mathematics seriously altogether.

(Bokulich and Maudlin are objecting specifically to naively reading the physical ontology off the mathematical formalism, which is not quite my suggestion. To say that a theory's mathematical structure is a guide to the world's physical structure is not yet to say anything about the ontology, about what entities instantiate the structure. We saw this by example in Chapter 5: being a realist about spatiotemporal structure, in my sense, is compatible with different views on the physical ontology. For all that I have said here, the ontology may well be something that we cannot read directly off the formalism. Although I did suggest something close to that in North (2013), I now think the issue is more subtle. For instance, as we saw in Chapter 4, there will typically be some initial assumptions made about the ontology, something I expand on in the next chapter. Better to say that the mathematical formulation of a theory and the physical ontology constrain each other.)

Any view that takes the mathematical structure of a physical theory seriously will face hard questions about how, and why, we should choose one mathematical formulation over another when different ones are available; and why, and when, we should infer physical structure in the world from one formalism, or from one aspect of a formalism, rather than another. But the fact that a view gives rise to hard questions is insufficient to impugn it. In any case, these are exactly the kinds of questions I have been addressing here.

As I have said before, this seems to me to be part and parcel of a basic realism about our fundamental physical theories. These theories are formulated in abstract mathematical terms, and the realist believes (roughly) that our best physical theories describe the world, they tell us what the physical world is really like. For the realist, then, the mathematical structures in which these theories are formulated must be telling us something about the physical world. As Wallace says, it seems all but impossible "to find some way in which a physical theory represents the world *other* than by some sort of correspondence between the mathematical description and the physical reality" (2012, 28; original italics). At the same time, this does not mean denying that some of our mathematical tools are just that—merely useful tools or descriptive devices—nor that the correspondence may be indirect to varying degrees.

You may nonetheless think that an emphasis on mathematical structure is particularly wrongheaded when it comes to *empirical* theories. Maudlin (2013) criticizes the method of extracting metaphysics from physics that begins with a standard mathematical formalism rather than the experimental facts on which any physical theory will be based. Yet that's not quite what is going on here. The realist about structure, in my sense, enters the conversation assuming that certain formulations have been chosen for good scientific reason. There is good reason, the realist thinks, for the way in which a theory has been formulated, above all the accumulated experimental evidence. That said, the experimental facts on their own will not pin down a single formalism (nor a single metaphysics to go with it: Chapter 7), and it is here that the realist about structure aims to figure out which of the available formulations to choose.

My approach may still seem to sanction a naive attitude toward candidate fundamental theories, one that aims "to interpret our physical theories by taking their mathematical structures at face value," as Barrett (2019, 1190–1) puts it. (The phrase comes from Maudlin, who argues against the "desire to take the mathematics at face value, insofar as possible, when proposing an ontology" for a physical theory (2013, 138). Maudlin is objecting to wavefunction realism in particular, the view that the mathematical wavefunction appearing in the formalism of quantum theory directly represents a real physical field.) According to the face value interpreter, we directly read off the physical nature of the world from a theory's mathematical formalism, simplemindedly taking the mathematical structure of a theory (or of the models of a theory, as Barrett also puts it) to directly represent the physical world. We unreflectingly "read the ontological implications of a theory off the formalism," in Bokulich's words, by way of "a facile identification of the formalism with the world" (2020, 186).

Although I do advocate a reasonably straightforward means of interpretation, we can distinguish the sort of face value interpretation that I support from an unreflecting kind that many with good reason would reject. This is because the realist about structure does not simply read off the metaphysics directly from the mathematics without any further ado.

First, as mentioned above, the realist about structure is not so naive as to read a genuine physical feature into every single mathematical feature of a formalism. There can be mathematical features that do not directly correspond to features of the physical world, and the realist about structure aims to figure out when this is the case. At the same time, the mathematical structure required to support the fundamental laws, such as the inertial structure presupposed by Newton's laws or the structure required to state the Lagrangian laws in terms of generalized coordinates, does tell us something about the world, and in such a way that it differs from what the mathematical structures required by other mathematical formulations tell us. Face value interpreters—or what might be better called "reasonably straightforward" or "sophisticated face value" interpreters—have to be careful whenever there are different mathematical structures we can use to formulate a theory. In that case, we will have to investigate what reasons there may be to choose one over another. Again, this is an admonition to be careful, not a reason to avoid taking the mathematics seriously as an epistemic guide altogether.

Second, a reasonably straightforward method of interpretation does not mean reading everything about the physical world off a bare formalism. Some interpretive assumptions will go into choosing a formalism to begin with. When investigating the mathematical structure required for Newton's laws, for example, it was important to have some initial physical assumptions, such as that of a fundamental ontology of point-particles interacting by means of forces. Without *some* initial conception of how a formalism describes the world and how it was devised on the basis of empirical evidence (a conception that can of course be revised and refined in light of further evidence), we would get nowhere in our interpretive project. Such assumptions are crucial to understanding how the theory is empirically confirmed, for starters. (It does not make much sense to take a bare formalism, some mathematical equations with no interpretive links to the physical world, as having been empirically confirmed.) In other words, we should not take a bare formalism in isolation and somehow read off the picture of the world entirely from it, as the phrase "face value interpretation" suggests. According to the properly sophisticated face value interpreter, that picture of the world plays a role in choosing the mathematical formulation in the first place—as in Newtonian mechanics, a metaphysics of forces suggests laws mathematically formulated in terms of vectors.

Any physical theory will require some initial physical postulates, some initial assumptions about what there is and what it is like, which allow us to get a grip on how the formalism connects to the world. These will already start to suggest that some mathematical formulations better or more perspicuously or more directly represent the world than others. (As an initial assumption that the world comprises n classical point-particles in three-dimensional physical space indicates that a three-dimensional mathematical space more directly represents the physical world than a $3n$-dimensional space does.) At the same time, those initial posits will not answer all the interpretive questions that arise about the formalism.

(Nor is a commitment to face value interpretation the entirety of what is going on behind wavefunction realism, the primary target of both Maudlin (2010, 2013) and Bokulich (2020). Albert, for example, a prominent advocate of the view, does not simply read the ontology off a given formalism. He does not think that statespace formulations of classical theories indicate that physical space, the space of ordinary material bodies, is anything other than three-dimensional, for example, whereas he does argue that an analogous conclusion is forced on us in quantum mechanics (Albert, 1996, 2019a). Albert (2015) gives other reasons for thinking that the mathematical wavefunction represents a real physical field, for instance that it evolves "in accord with a dynamics that seems to present the various adjacent pieces of it as constantly *pushing* and *pulling* on one another—just as the various adjacent pieces of gravitational or electromagnetic fields do," so that it has "every characteristic sign and signature of concrete mechanical *stuff*" (126; original italics). Alyssa Ney (forthcoming) discusses a variety of reasons behind the view, above all the desire to have a fundamentally local and separable metaphysics. Taking the mathematical wavefunction seriously as directly representing the physical ontology is not based solely on a naive reading of the metaphysics directly from the mathematics, in other words: there are further considerations that go into the view.)

A third reason the face value interpreter needn't be so naive is that our inferences from the mathematical formalism to the nature of the physical world needn't occur by way of a simple or direct or complete isomorphism between the world and the formalism, an exact match in all respects: the representation relation can be more indirect than that. Classical mechanics is a case in point. Any classical theory can be formulated in terms of its statespace. The statespace formulation captures all the relevant physical facts (so that we can learn about the nature of a classical mechanical world from it), but not by means of a direct match between all aspects of the formalism and features of the physical world. Given an initial physical posit that the theory is about n particles in three-dimensional space, it is clear that the statespace formulation represents the physical world somewhat indirectly.

(Relative to a different initial posit, the high-dimensional representation may be more direct: Chapter 7.)

In physics, we often use mathematical objects that have a different surface structure from the physical things they represent—a classical mechanical world does not itself consist of a single particle moving through a high-dimensional physical space, even though it can be represented by a single point tracing out a trajectory in a high-dimensional mathematical space; a quantum mechanical world does not itself consist of a ray evolving in Hilbert space, even though it can be mathematically represented in this way—but the representation relation is there nonetheless. These mathematical objects do represent, and can therefore tell us about, the physical world, albeit in many cases somewhat indirectly.

For all these reasons, "face value interpretation" is a misleading characterization of what is going on. The underlying idea is simply to take the mathematical formulations of our best physical theories seriously in telling us about the nature of the physical world, an idea that any realist should agree with. Such a realism about structure starts from the utterly benign claim that in physics, we use mathematical objects to represent the physical world; that, "Mathematical structures are used in mathematical physics as *representations* of the physical world" (Maudlin, 2014b, 796; original italics). This will of course lead to difficult questions about which features of the mathematics are genuinely representational and which are not; how the representation relation works for those features that are; and to what extent different types of mathematical features represent the world. I have argued that there are a few key principles that guide our inferences from the mathematics to the nature of the physical world, principles that we are justified in adhering to, but I do not claim to have given a wholly rule-bound and precise procedure. (Scientific theorizing is too messy for that.) Regardless, the mathematical formalism required by our best physical theories surely tells us *something* about physical reality, so that it should be an epistemic guide to the nature of that reality. This much should be uncontroversial, at least to the realist. For that matter, it should be uncontroversial even to certain antirealists, like Ruetsche or van Fraassen, who take seriously what such a theory is telling us about the physical world (while refusing to believe what it is telling us).

6.3 A different realism about structure

Wallace and Timpson (2010), as well as Wallace (2012), suggest a different, in some ways weaker—though in other ways, stronger—realism about structure. They take a theory's mathematical structure seriously as a guide to the nature of physical reality. But their view leaves room to avoid seeing any genuine differences between theoretical formulations ordinarily taken to be equivalent, such as the two formulations of classical mechanics discussed in Chapter 4.

In one respect, Wallace and Timpson take a theory's mathematical structure *more* seriously than I do, for they take it to be the sole guide to the physical ontology. They say that, "[I]n our view, there is no guide to the ontology of a mathematically formulated theory beyond the mathematical structure of that theory" (Wallace and Timpson, 2010, 702). At the same time, they remain agnostic about whether, when there are different mathematical formulations that appear to describe physical reality equally well, one of them must be closest to the truth about that reality. They suggest that there can be different formulations that equally describe physical reality, without its being the case that one of them is most accurate. (Wallace and Timpson (2010) say they wish to remain agnostic about the possibility; Wallace (2012) (especially Sec. 8.8) seems to positively endorse it.) This is the respect in which their view is a weaker realism about structure: there is more room for maintaining that allegedly equivalent theoretical formulations, such as Lagrangian and Newtonian mechanics, are equally accurate descriptions of the world, as usually thought.

Wallace and Timpson's aim is to defend a particular understanding of quantum theory, opposed to wavefunction realism. In defending their view, they draw an analogy to statespace and ordinary-space formulations of classical mechanics. They note that any theory of classical mechanics can be formulated in two different, mathematically equivalent ways: in terms of n particles in a three-dimensional space or a single particle in a $6n$-dimensional space. Each of these formulations provides a legitimate, accurate representation of classical mechanical reality. However, Wallace and Timpson say, the former representation is more "perspicuous" or "sensible." This is because, even though the single-particle, high-dimensional description may not be any further from the truth—it represents classical mechanical reality entirely accurately—it is harder on the basis of this description to discern the nature of the reality being represented. Wallace and Timpson conclude that neither formulation is closer to the truth about the physical world, even though one of them is better than the other—not in the sense of being closer to the truth, but in the sense of being more perspicuous.

Similarly, they say, for quantum mechanics. Thinking about quantum mechanical reality in terms of four-dimensional spacetime, in line with a view they call "spacetime state realism," rather than in terms of the extremely high-dimensional space of the wavefunction (as per the wavefunction realist), yields the most perspicuous understanding of quantum theory. This is the case even though (they suggest, and Wallace (2012) argues) physical reality itself is nothing but the quantum state: physical reality is, or directly corresponds to, the wavefunction evolving linearly and deterministically, mathematically represented by a vector (or ray) evolving in a high-dimensional Hilbert space. Thinking about the quantum state in terms of properties assigned to regions in ordinary spacetime yields the best understanding of quantum reality, in the same way that a description in terms of n particles in three-dimensional space yields the best understanding of classical mechanical reality. In particular, the opposing wavefunction-realist view, according to which quantum reality is fundamentally a complex-valued field on (a space isomorphic to) $3n$-dimensional configuration space, yields a highly imperspicuous understanding of the quantum state, for it obscures various features of it, in the same way that a description in terms of a single particle in $6n$-dimensional space obscures various features of classical mechanical reality.

In their view, then, wavefunction realism is an equally legitimate description of quantum reality; it is simply "an *unhelpful* way to think about the ontology of quantum mechanics" since it "obscures the structure of the state" (Wallace, 2012, 317; my italics). The wavefunction-realist description of physical reality is not false. Yet just as with a multi-particle versus single-particle formulation of classical mechanics, spacetime state realism yields a much more perspicuous understanding of the theory and what it says about the world.

Notice there is something of a tension here. On the one hand, Wallace and Timpson want to say that wavefunction realism is an equally allowable, equally accurate, just as good description of physical reality, in the same way that a high-dimensional formulation of classical mechanics is. On the other hand, their arguments against wavefunction realism (at least those of Wallace (2012)[1]) suggest that wavefunction realism is *not* an equally good description of quantum reality, since it "*mis*represents the structure of quantum mechanics" (Wallace, 2012, 316; my italics) (in particular, by singling out a preferred basis and by not accurately representing certain features of quantum field theory).

Regardless of what Wallace and Timpson ultimately want to say about wavefunction realism—that it is just as legitimate a description of quantum reality as spacetime state realism is, or that it misdescribes certain things and is in that sense not equally legitimate—it is helpful to see their approach in the way that Sider (2020, Ch. 5) characterizes it, as what he calls a "quotienting" view. The formulation of any physical theory typically contains some aspects that are merely conventional.[2] Further, there will typically be different theories, descriptions, or formulations that are all equally good representations of the physics in that they disagree solely on such conventional matters. The quotienter thinks that, whenever this is the case, we needn't say any more about the nature of the world other than to give the different theories or descriptions and state that they are all equally good. We needn't say what it is about the world that makes them all equally good, that is: we may simply stipulate that they are equally good. If you then ask what reality

[1] Wallace and Timpson (2010, 701) state that they aim to be neutral on this: "while we argue for the adoption of spacetime state realism over wave-function realism, we wish to remain neutral on whether one of these (or perhaps some third) really does provide the One True Interpretation of the quantum state, or whether one is merely a more perspicuous description than the other, a description of something that we are ultimately unable to render unequivocally in intuitive terms."

[2] Sider describes the quotienting view as an attitude one can have toward any theory of the world, including broadly metaphysical theories. My focus is on physical theories.

is like in virtue of which the different descriptions do equally well, the quotienter's answer will be simply that "reality is such as to be well described in any one of the equivalent ways" (Sider, 2020, 192). In other words, the best theory or description of the world can be given by an equivalence class of theories, where we do not say why that is the relevant equivalence relation. In giving the best theory, we can simply "quotient out" the conventional content "by hand."[3]

Wallace and Timpson's view can be understood in these terms. They allow that there can be different, equally legitimate descriptions of physical reality, as they say of both classical and quantum mechanical reality, even though there may be no more to say about *why* the different descriptions are equally legitimate, as they suggest of quantum theory in particular. One description may be more perspicuous than the others, but we do not—perhaps cannot—say what it is about the world in virtue of which it is most perspicuous.

This kind of realism about structure can avoid the more controversial conclusions of Chapter 4 (as well as those to come in Chapter 7). Although Wallace and Timpson think that a theory's mathematical structure tells us about physical reality, they do not suggest that differences in mathematical structure, in my sense, must correspond to differences in physical reality. This leaves room for them to agree with me that in certain cases there are such mathematical differences, while denying that this must amount to any physical difference. Lagrangian and Newtonian mechanics, in particular, could both fall within the equivalence class of equally good theories. If someone then asks: what is it about the world that makes these equally good descriptions?, Wallace and Timpson can simply say that they are equally good, and leave it at that. They do not have to say that, nor in what way, a world built up out of fundamental vector quantities is different from one built up out of scalar energy functions, say, instead stipulating that these are equally good representations of a classical mechanical world. (Although they note that ordinary-space and statespace formulations of classical mechanics are isomorphic by construction, they may allow for the possibility that non-isomorphic theoretical formulations can be equally good descriptions of physical reality, as is arguably the case for spacetime and configuration-space formulations of quantum mechanics. If it turns out that they do not want to allow for this, then whether they agree with my conclusions about classical mechanics will depend on whether they agree that there are any structural differences between the formulations.)

As Sider notes, a quotienting view adopts a more conventionalist attitude toward metaphysics and ontology than does standard realism, and we can see this of Wallace and Timpson's view in particular. According to Wallace and Timpson, different formulations of classical mechanics that superficially seem to disagree about whether reality is fundamentally $6n$- or three-dimensional, for example,

[3] See Dewar (2015, 2019) for discussion of another view that is like quotienting.

do not really disagree, but are simply different, equally legitimate descriptions of the same physical reality. The standard realist will largely agree, but will also think there is something more to be said about why one of these descriptions is more perspicuous (more below). Wallace and Timpson's claim about the different descriptions of quantum reality in particular is not a standard conclusion, at least not among the realists who debate the question of wavefunction realism. According to Wallace and Timpson, the description in terms of four-dimensional spacetime and the one in terms of an extremely high-dimensional space are equally accurate depictions of quantum reality (even if one is more perspicuous). The two descriptions do not really disagree, and the dispute over wavefunction realism is not a genuine or substantive dispute.

This is a more deflationary attitude about the metaphysics of quantum theory than that of your typical realist. It is not a full-fledged conventionalism about metaphysics and ontology: *there is* an ontology and structure in the world; there are objective facts about these things, which the different descriptions all equally represent. There is nonetheless a conventionalist element in the denial that we can say what those facts are beyond giving the relevant equivalence class of theories. For that reason, I find the view insufficiently realist. (Sider argues that it is unsatisfying; I agree, but want to emphasize the way in which it is particularly unsatisfying to the scientific realist.) The view is realist in that there is a description-independent reality that physics is getting at; yet physics is only able to describe that reality "at one level removed," by means of an equivalence class of theories, so that we never quite come out and say what reality is really like. We only say that physical reality is such as to be well-described in any of these ways. That is like giving a theory of Newtonian mechanics which states that reality is such as to be accurately described in terms of the coordinates of any inertial frame, without going on to say what it is in virtue of which these all yield accurate descriptions. This is an unsatisfying endpoint. The realist wants to know: what is it about reality that makes the description in terms of any inertial frame equally good? Without an answer to this question, the view feels uncomfortably close to Bohr's idea that physics is not about nature and what it is like, but only about what we can say about nature; that physics is about our theories or descriptions of the world, rather than the world itself.

This won't move Wallace and Timpson. In his book, Wallace notes that the "metaphysical problem [for the Everettian in particular] is the problem of what this 'quantum state', which is supposed to represent the whole of microphysical reality, actually *does* represent," but he goes on to suggest that this problem needn't be taken seriously on the grounds that, "Physics is ultimately concerned with making claims about the *structure* of the world, described mathematically, rather than its 'true nature'" (2012, 42; original italics). To me, this comes too close to giving up on the realist project altogether. The realist wants to understand what the world is like (its true nature, if you like), given its best physics. And it is simply not clear what the world is like, on Wallace and Timpson's view, given the extent of the seeming differences between formulations that are, in their view, equally accurate. By way of comparison, Wallace and Timpson point out that we do not worry about the lack of an intuitive understanding of the electromagnetic field: we are satisfied with seeing it as an assignment of certain properties to spacetime regions, without saying anything more about its nature; and they suggest that we understand the quantum state analogously. But the worry is not about the lack of a particularly intuitive picture of the quantum ontology. The worry is that there is no positive picture of the ontology at all, if the most we can do is to give an equivalence class of theories that appear to radically differ in what they say about the structure and ontology of the world. As Allori puts it, "If the ontology of a theory is not clear, then it is not clear what entities the theory is assuming to exist," which in turn makes it unclear how we can embark on the "enterprise of inferring what the world is like" from the theory (2015b, 108). On Wallace and Timpson's view, that enterprise is not clear.

Their view is clearer in the case of classical mechanics, with good reason. In this case, we are unbothered by there being two theoretical formulations that are both accurate. The reason is that the formulations are stipulated from the beginning to both be describing a physical reality of particles moving around in three-dimensional physical space. It is clear from the outset what the physical reality *is* that the two descriptions agree on (even if certain features, such as the structure of physical space, are not evident from the beginning but require further interpretive work). There is a clear physical picture of the world that the two descriptions have in common, which we can point to as the reason for our judgment of equivalence, as well as our judgment that one of them is more perspicuous (which might further lead us to say that one of them more directly or more accurately gets at the nature of physical reality). This is not the case in quantum theory, where the question of what the different descriptions represent, the initial physical postulates of the theory, has not been settled at the outset. That is why it is so hard to discern what the nature of the reality is of which the different descriptions are all alleged to be equally accurate.

One might go so far as to say that the quotienter does not even offer a complete fundamental physical theory. Maudlin says the following of the postulates of quantum mechanics one finds in standard textbooks: "What the quantum recipe does not resolve, what it does not even purport to address, is *what the physical world is like such that the quantum recipe works so well.* To answer this question, we need . . . a physical theory, a clear specification of what there is in the physical world and how it behaves" (2019, 77; original italics). The quotienter arguably does not answer that question, and therefore lacks the "clear ontology" required of any fundamental physical theory (Allori, 2015b, 108). I won't take a stand on this here—I won't discuss what it takes to be a sufficiently clear or "precisely articulated physical theor[y]" (Maudlin, 2019, 94)—but it does seem to me that the history of foundations of quantum mechanics has taught us that the realist ought to be moved by such a thought.

Wallace and Timpson's realism about structure is both stronger and weaker than mine. It is weaker in that it does not require an explicit picture of what the world is like beyond offering an equivalence class of theories, which leaves room for saying that Newtonian and Lagrangian mechanics, for example, are merely alternative descriptions of the same fundamental physical reality. At the same time, it is stronger precisely because it takes a theory's mathematical structure to be the *sole* guide to the ontology or the metaphysics more broadly. In this respect, the view's method of interpretation is even more "face value" than my own: there is nothing more to say about the nature of the world other than what can be read off the formalism.

Although the quotienting view strikes me as going against the tenor of scientific realism, it is worth noting that this type of view is not unusual among philosophers of physics these days, even among those who claim to be realists. Consider James Weatherall's argument (2016a) that traditional Newtonian gravitation and geometrized Newtonian gravitation—the latter seeming to "geometrize away gravity" in the manner of general relativity—are equivalent in an important sense, namely empirically and structurally (in particular, categorically) equivalent, given a natural formulation of the traditional theory. (More on this pair of theories in Chapter 7.) These theories might seem to disagree on such things as whether there are gravitational forces and whether spacetime can be curved. Yet according to Weatherall, we should not conclude that they really do say anything physically different, given their empirical and structural equivalence; rather, they "say the same things about the world" (2016a, 1087): they are wholly equivalent, different formulations of one and the same theory. More generally, "theories that attribute apparently distinct geometrical properties to the world . . . may provide different, but equally good, ways of representing the *same* structure in the world" (Weatherall, 2016a, 1086; original italics).

This can be seen as a quotienting view. Different theoretical descriptions that appear to disagree on such basic features of the world as the structure of spacetime and the existence of certain forces are taken to be equally good descriptions of the same physical reality. The view is realist in that *there is* some structure and ontology which the theories are equally good at describing; there are objective facts about these things. But we do not say any more about *why* these are equally good descriptions, beyond demonstrating their formal and empirical equivalence and stipulating that they are therefore equivalent through and through. (Although Weatherall does not put his view in quotienting terms, it is hard to make sense of his claim that the theories attribute the "same structure" to the world absent the quotienting idea. He does not say what the structure is that the two theories agree on, beyond the fact that it is equally well-represented by means of either formalism.)

The standard or traditional scientific realist won't be content with this. The standard realist will want to know: assuming that our scientific theories tell us about the world, how can these be equally good representations? What is it—what could it be—about the spacetime geometry that makes them equally good representations? Weatherall's view does not answer this question. (Going too far in the other direction may lead you to worry that we will find too many cases of inequivalent theoretical descriptions; I discuss this in Chapter 7.)

It is worth emphasizing the extent to which the quotienting view diverges from scientific realism as traditionally understood. Realism is supposed to be the view that, unlike instrumentalism, say, explains why our scientific theories are as successful as they are. The quotienter does not give an explicit answer to this question, which leaves the view unsatisfying in the same way that instrumentalism is unsatisfying. As Jones puts it, the realist believes that science progresses toward "an ontologically well-defined world picture": the realist "envisions mature science as populating the world with a clearly defined and described set of objects, properties, and processes, and progressing by steady refinement of the descriptions and consequent clarification of the referential taxonomy to a full-blown correspondence with the natural order" (1991, 185–6). (He says this in the context of arguing that various theories of contemporary physics create trouble for this vision.) The quotienter never realizes this "full-blown correspondence" or "ontologically well-defined world picture." There is "a nature of things itself," as Jones puts it, and in that sense the view is realist. But the quotienter does not go on to say what, exactly, that nature of things is. In this respect the view fails to fully realize the realist vision.

There may be a more nuanced quotienting view that does not come so close to instrumentalism. Consider a reasonably standard idea among contemporary spacetime substantivalists that in general relativity, an equivalence class of diffeomorphically-related solutions to, or models of, the field equations represents a single physical possibility. This can be seen as a quotienting view that does not broach instrumentalism: plenty of realist philosophers of physics take this to be the right way to think about spacetime in relation to the models of general relativity (in particular in the face of the hole argument: Chapter 5, note 13). Perhaps the problem with the above views is not quotienting per se, but a too-coarsely applied quotienting rule. In the current case, the equivalence relation seems to truly "quotient nature at the joints."[4]

Even here, however, a more thoroughgoing realist will want more than just a brute stipulation that this is the relevant equivalence relation, but also an explanation for why it is—an account of what the fundamental physical objects and properties and relations are, and of the invariances in the laws, that make this the right equivalence relation, the one that does carve at the joints. In refusing to offer such a thing, even the more sophisticated quotienting view retains an unsatisfying air, if not of instrumentalism, at least of a not fully-fledged realism, in Jones' sense, about it.

[4] Thanks to Laura Ruetsche for the suggestion and the phrase.

Finally, it is hard to reconcile the quotienting view with the claim that some of the equally allowable descriptions are more perspicuous than others, as Wallace and Timpson say. It is hard to see how to maintain that one description is more perspicuous than another, without also saying that one description *better* represents physical reality than the other—that one is *closer to the truth* than the other, and that is *why* it is more perspicuous. The standard realist says that the three-dimensional formulation of classical mechanics is more perspicuous *because there are n* particles moving around in three-dimensional space, for example. This description gets closer to the truth about the physical world; it is more direct; that is why it is more perspicuous. Wallace and Timpson might agree with this for the case of classical mechanics. But they withhold from saying anything like this for quantum theory, which leaves one wondering exactly what the claim that one description is most perspicuous amounts to. What are the different accurate formulations representations *of*, such that we rightly deem one of them a particularly perspicuous representation?

It is the realist inclination to want to understand why one description of physical reality is better or more perspicuous than another. The realist wants a perspicuous theory, with an ontologically well-defined world picture, which is perspicuous because it best corresponds to the nature of reality. (This might spur us to seek an entirely convention-independent description, which most directly represents the structure of the world. I am sympathetic to such a project, but do not take this on here.) It is the realist hope that physics need not give up on the project of explicitly specifying what the nature of reality is. The quotienting view does not get us this. For some, that may be enough. For me, it is too weak-hearted a realism. It is worth aiming for more.

6.4 Structure, models, and scientific theories

One kind of inference I have been discussing concerns the structure of a given theory and what it tells us about the world. Another kind is inter-theoretic: given two physical theories and the structures they require, what can we infer about the relationship between the theories and the worlds they describe? That was the focus of Chapter 4.

It may seem as though my realism about structure commits me to a particular answer to the second question, alongside a particular criterion for the equivalence of physical theories. Halvorson (2012) and Barrett (2019) say that I am committed to what they call the *model isomorphism criterion*, according to which, "Theories T_1 and T_2 are equivalent . . . if for every model of T_1 there is an isomorphic model of T_2, and vice versa" (Barrett, 2019, 1170).[5] But as Halvorson and Barrett both argue,

[5] See also Barrett (2020b). They say that the argument in North (2009) for the non-equivalence of Lagrangian and Hamiltonian mechanics in particular relies on this. Similar things could be thought of the discussion in this book, especially Chapter 4.

this cannot be the right criterion for theoretical equivalence. (For our purposes, we may think of a model of a theory as some sort of abstract mathematical representation of a possible world of the theory.)

Consider non-relativistic quantum mechanics given in terms of Heisenberg's matrix mechanics versus Schrödinger's wave mechanics. These formulations are mathematically and empirically equivalent. (I do not mean the actual theories presented by Heisenberg and Schrödinger—which F. A. Muller (1997a,b) argues were not equivalent, neither mathematically nor empirically, at the time they were developed—but the formulations that currently go by the names, often called the Schrödinger and Heisenberg "pictures.") The standard view in physics is that these are the same theory, differently formulated—a mere difference in notation. However, as Halvorson and Barrett point out, the model isomorphism criterion deems them inequivalent formulations, and hence distinct theories. A matrix algebra (the mathematical formalism of the Heisenberg picture) cannot be isomorphic to a space of wavefunctions (the formalism of the Schrödinger picture), for the simple reason that these are different types of mathematical object, and the mathematical notion of isomorphism only applies to mathematical objects of the same type.

Or consider general relativity formulated in terms of a manifold with metric of signature $(3, 1)$ versus $(1, 3)$. The two formulations differ in only a conventional choice of sign. However, as Barrett (2015a, 2019, 2020b) notes, models with metrics of opposite signature cannot be isomorphic, simply because metrics of different signatures cannot be isomorphic. The metric functions will assign opposite signs to the length along a continuous path between two spacelike or timelike separated points, and so different "distances" between them. The model isomorphism criterion deems the formulations inequivalent—an unacceptable result: this is a paradigmatic case of a mere notational difference.

Examples such as these more generally seem to spell trouble for the kind of realism about structure I am advocating. The given pairs of formulations are notational variants, but nonetheless differ in mathematical structure in a sense. Once again, it seems as though I am taking theories' mathematical structures *too* seriously.

The model isomorphism criterion may seem to lie behind my approach in particular when combined with a few other ideas: the semantic conception of scientific theories, according to which a theory is identified with its set (or class) of models (plus a "theoretical hypothesis" outlining how a model represents the world); so that a theory's mathematical structure is just the structure of its set (or class) of models; and the equivalence of physical theories is a matter of whether the theories have pairwise isomorphic models.

However, my view does not require any of these ideas, and it is not committed to the results being claimed. To the extent that it does support conclusions in the vicinity, they are not so untoward as all that. There are a few reasons for this, which will start to suggest some general lessons for the equivalence of physical theories, the topic of the next chapter. (You may already be convinced that I do not rely on the model isomorphism criterion; there will also be some general points made that serve both to distance my approach from others in recent literature and to highlight certain ways in which it is in agreement with them.)

First and foremost, the model isomorphism criterion relies on a different sense or aspect of a theory's structure from what I do. In Chapter 4, I concluded that Lagrangian and Newtonian mechanics are not equivalent because they differ in the structure required by their dynamical laws, as evidenced alternately by the structure presupposed by the equations that represent the laws, and the structure possessed by the statespace on which those equations can be defined.[6] The model isomorphism criterion says that two theories are inequivalent when the structures of their individual models do not align in the right way: when there is a model of one theory for which there is no isomorphic model in the other. My argument for the inequivalence of different formulations of classical mechanics does not rely on that criterion for the simple reason that a theory's dynamical or statespace structure is not present within any one model of a theory; *ipso facto* it is not present within any pair of models consisting of one model taken from each of two different theories. It is rather a structure on a theory's set of models, if one wants to put it that way. (In Chapter 4, we saw some reasons for not wanting to put things in that way.)

This is to endorse an idea that Halvorson suggests in the course of arguing against the model isomorphism criterion, which is that "theoretical equivalence is global" (2012, Sec. 4.2). The equivalence of theories—that is, their formal or structural equivalence (reasons for the precisification in the next chapter)—concerns the structures of theories as a whole, something that isn't revealed by a pairwise comparison of theories' models. I also endorse a global view of the structural equivalence of theories, which amounts to rejecting the model isomorphism criterion.

Indeed, not only is the relevant structure not present within any single model, but neither is it present within the entire collection—that is, the bare set (or class)—of a theory's models. Think of it in terms of the statespace. Each point in a theory's statespace corresponds to a different possible world or model of the theory, but the statespace as a whole is not just the set of those points. There is some further structure to the statespace, over and above the set structure of the collection of worlds or models taken together, which represents how the different possible worlds or models are related to one another.

This is to endorse another idea of Halvorson's (2016), which is that scientific theories are not "flat" but have structure (a point that seems to have gotten lost

[6] North (2009) concludes that the Hamiltonian and Lagrangian formulations of classical mechanics are not equivalent for similar reasons.

in the usual debate over the semantic versus syntactic conception of theories). Scientific theories are not just bare collections of models (as the semantic conception says) or sentences (the syntactic conception): they also specify relations that hold between different models or sentences.[7] Our best scientific theories tell us not only which behaviors, worlds, or histories are possible, but also which ones are possible given various alterations to initial conditions. They specify more than a bare set of possibilities, but also relationships between different possibilities, such as which ones are "closer," or more similar to, or more easily obtainable from, which others. This is crucial to the backing of the counterfactual claims that are central to theories' explanations and predictions of the phenomena. Whatever a scientific theory is (I won't take a stand on this here), it seems to me that it must be a structured thing in Halvorson's sense. The model isomorphism criterion simply ignores this structured aspect of theories.

And once we reject the model isomorphism criterion, it is clear that we don't have to say that the above two formulations of general relativity are inequivalent. Indeed, by my conception, they are equivalent in terms of structure, for the very reason that they differ in only a conventional choice of sign: this is exactly the kind of thing that does not amount to a difference in structure in my sense. It does not matter whether timelike vectors get negative or positive lengths (so long as they have lengths of opposite sign to spacelike vectors): the physics tells us that this is an arbitrary difference in description; the laws say the same thing regardless. (The two ways of assigning lengths to spacetime vectors play equivalent roles in each formulation, even though their models are not pairwise isomorphic: a global equivalence in Halvorson's sense.) Compare: we could define the metric of a three-dimensional Euclidean space to have signature $(-, -, -)$, yielding a space that is strictly speaking non-isomorphic to an otherwise-similar space of metric signature $(+, +, +)$. Yet we surely want to be able to say that spaces with these metrics have the same structure or geometry, the only difference being a conventional choice of sign.[8] The model isomorphism criterion does not get this result; my conception of structure does.

(There is arguably a possible physics that does treat metrics with different signatures differently, whose laws pay attention to the sign given to different lengths. An analogy. Take a physical theory whose laws presuppose a temporal orientation, a global structural distinction between the two temporal directions, and contrast it with a theory whose laws additionally assume an objective fact as to which direction is past and which is future. That is, compare a physics that

[7] Halvorson (2016) describes how each of the traditional "flat" views can be straightforwardly modified to become a "structured" view. He further suggests that category theory is a natural framework for developing this conception of scientific theories as structured entities.

[8] As one book says: "If the metric is positive-definite, then its canonical form must have all +1s, and the space is Euclidean. If the metric is negative-definite it is also said to be Euclidean, since what is important for the space is whether the signs are all the same or not" (Schutz, 1980, 66).

presupposes a temporal orientation with one that also says which of the two globally definable temporal orientations is the right one. According to the first kind of physics, there is an asymmetric temporal structure, a genuine difference between the two temporal directions, but there is no objective distinction between a world and its time reverse: it is arbitrary which of the two temporal orientations we choose, although the physics requires there to be an orientation. According to the second kind of physics, there is in addition a fact about which temporal orientation describes things correctly. A theory that is not symmetric under time reversal is an example of the first kind of physics.[9] A theory that posits a passage of time in the sense of Maudlin (2007b) is of the second kind. According to the second kind of physics, a world with one temporal orientation and a world with the other are objectively different worlds—just as a world with metric signature $(1, 3)$ is objectively different from a world with metric signature $(3, 1)$ according to the physics we are imagining.)

Similarly for the different formulations of quantum mechanics. A matrix algebra is not mathematically isomorphic to a space of wavefunctions in that these are different kinds of mathematical objects. However, these both instantiate a Hilbert space structure. They are isomorphic *qua* Hilbert space structure, exhibiting a similar structure at this more abstract level (which corresponds to their statespace structure). So that even if a "simpleminded isomorphism criterion" (Halvorson, 2012, 188) deems them structurally inequivalent, there is room for a less simpleminded isomorphism criterion, which agrees with the standard thought that these are mathematically or structurally equivalent, isomorphic, in a relevant sense.[10] (One might nonetheless, and more controversially, reject the usual view that these are wholly equivalent theories, for reasons discussed in Chapter 7.)

A related, more general lesson of these examples. The model isomorphism criterion requires that the mapping we use to draw conclusions about equivalence apply at the level of theories' models. Yet as we can see from the above cases, there are reasonable such conclusions to be made on the basis of a more flexible conception of the kinds of objects that can be related by the isomorphism, and of the structures they have relative to which the correspondence is an isomorphism, a mapping on the basis of which we may conclude that there is a relevant equivalence. To put it a slightly different way, we don't insist on specifying the relevant notion of isomorphism in advance (neither which collections of objects must be related by the mapping, nor what structures they have relative to which the correspondence is an isomorphism): there will be different isomorphisms that one can define, and it remains an open philosophical question whether a given one indicates a pertinent sense of equivalence.

[9] Earman's (2002) discussion of time reversal suggests a distinction between these two types of physics; see especially pp. 257–8.

[10] See Ruetsche (2011, Ch. 2) for discussion of this kind of equivalence and its physical significance.

Another illustration of this general lesson. Nicholas Teh and Dimitris Tsementzis (2017) say that the idea in North (2009) (shared by Curiel (2014)) that "Hamiltonian mechanics is not isomorphic to Lagrangian mechanics" is puzzling; for "any attempt to resolve the question of whether Hamiltonian mechanics can be isomorphic to Lagrangian mechanics must either (i) exhibit an appropriate category [or type of structure] in which 'isomorphism' is understood, or (ii) show that no such category exists" (2017, 46), neither of which has been done. They go on to locate a structure with respect to which one can define an isomorphism, and conclude that the theories are equivalent, as usually claimed, albeit for somewhat different reasons. At the same time, however, they also note that the theories' respective statespaces, with the relevant scalar function, "are very different structures" (Teh and Tsementzis, 2017, 47). And it is for this very reason that we are justified in concluding that the theories are not structurally equivalent, not isomorphic, in the sense that they employ different statespace structures. To my mind, this is moreover a physically significant respect in which the theories are inequivalent, even if they are structurally equivalent at the extremely abstract level that Teh and Tsementzis demonstrate. (Recall the point from earlier chapters that an isomorphism depends on the type of mathematical structure in question, with the result that two objects can be isomorphic in some respects or with respect to some structure, while being non-isomorphic in other ways or with respect to other structure. In this case, the two theories are not isomorphic in a relevant sense, despite their being isomorphic in other respects.[11])

(Concerns about taking the mathematical notion of isomorphism to be the standard for "sameness of structure" have moved many philosophers recently toward a category-theoretic understanding of structure and equivalence, and of scientific theories themselves, where what is relevant is the equivalence of categories, investigated via a kind of mapping called a functor, with an isomorphism being limited to the objects within a category.[12] I think the general notion of the structure required by the laws that I have been relying on suffices for my purposes, although it may be that a category-theoretic account ultimately does a better job. I won't address that here.)

Halvorson (2012) suggests that I am also committed to the semantic view of scientific theories. This may be assumed by standard structural realists, in the sense mentioned in Chapter 1. However, my own realism about structure is very different. And for the above reasons, I do not think that a scientific theory must be

[11] Using category-theoretic methods, Barrett (2019) shows that there is a way of specifying the theories on which they come out as structurally equivalent, and a way on which they do not, so that whether we conclude that the theories are equivalent will depend on what mathematical structures we take to be essential to them: recall Chapter 4, note 50.

[12] See especially Halvorson (2012, 2016) and Weatherall (2016a,b, 2017); also Teh and Tsementzis (2017); Halvorson and Tsementzis (2017); Barrett (2019, 2020b); Nguyen et al. (2020).

identified with its set or class of models, nor that its structure should be identified with the structure of its set or class of models.

In all, my notions of structure and sameness of structure do not align with those assumed by the model isomorphism criterion in particular or the semantic conception of theories in general. Nor do they require a model-theoretic conception of structure more generally. You might want to conceive of a theory's structure in terms of its models, and construe sameness of structure in terms of the structure of a theory's models in relation to that of other theories' models (whether by means of an isomorphism, or an equivalence between the categories of models, or some other notion). But this is not the only way of doing so, and the discussion in earlier chapters reveals that it is not how I am in general conceiving of things. The arguments here (as well as in North (2009)) do not rely on, nor do they even mention, the structures of theories' models.

There is a last, general reason to reject the model isomorphism criterion, which is that a structural equivalence is at best a necessary condition on the equivalence of physical theories, not a sufficient one as the model isomorphism criterion suggests.[13] Although I do think that an inequivalence in mathematical structure (of a certain sort) between physical theories suffices to indicate that they are not fully equivalent, I do not think that an equivalence of mathematical structure suffices for theories' wholesale equivalence. The simple reason is that a physical theory itself consists of more than just its formal apparatus or mathematical structure. So that even if two such theories are structurally equivalent in a relevant sense, there can be other, physically significant respects in which they fail to be equivalent. That is going to be the focus of the next chapter.

[13] Halvorson (2012, 187) and Barrett (2020b, 1185) state it as a necessary condition, while still arguing against it.

7
On the Equivalence of Physical Theories

> A doubt which makes an impression on our mind cannot be removed by calling it metaphysical; every thoughtful mind as such has needs which scientific men are accustomed to denote as metaphysical.
>
> Heinrich Hertz (1899, 23)

> I had spent six years slugging my way through many dozens of physics textbooks that were carefully written with the best of pedagogical plans, but there was something missing. Physics is the most interesting subject in the world because it is about how the world works, and yet the textbooks had been thoroughly wrung of any connection with the real world. The fun was missing.
>
> Jearl Walker et al. (2014, xv)

7.1 Differing criteria

I have been focusing on the mathematical structures in which our physical theories are formulated, and yet there is more to a physical theory than its mathematical formalism. The current chapter will underscore this point.

There has recently been increased attention paid among philosophers, especially among philosophers of physics, to the notion of theoretical equivalence: of when two theories say the same things in different ways, as we would say of a theory written in English as opposed to French, for instance. Philosophers have been trying to clarify what we mean when we say that two scientific theories are fully equivalent, or mere notational variants, and what criteria lie behind reasonable judgments of equivalence in physics.

Recent discussions have focused on various formal accounts of theoretical equivalence. Formal accounts say that scientific theories are equivalent when they are formally or mathematically or structurally equivalent in the right way (whether along the lines of a model isomorphism criterion, or a categorical equivalence, or a definitional equivalence, or some other formal criterion). Though referred to as "formal accounts," they typically require that the theories be empirically equivalent as well: the theories must make all the same observable predictions in some sense. This is included as a further constraint since it is generally thought both that empirical equivalence is a minimal condition for theoretical equivalence,

Physics, Structure, and Reality. Jill North, Oxford University Press (2021). © Jill North.
DOI: 10.1093/oso/9780192894106.003.0007

and that a theory's empirical content goes beyond its purely formal features.[1] (I won't try to analyze exactly what we should mean by empirical equivalence. It will be clear enough why the pairs of theories I discuss in this chapter are thought to be empirically equivalent.) Recent philosophical discussions compare and contrast different notions of formal equivalence, and go on to argue that one of these is the right such notion, the one that tells us when two physical theories are genuinely equivalent.

It seems to me these discussions have duly, but also *un*duly, emphasized the formal aspects of scientific theories. Something has been missing from the discussion, which is sufficient acknowledgment of the role that a theory's "picture of the world" will invariably play in our judgments of equivalence in physics. Recent discussions have been overly focused on the formal aspects of scientific theories at the expense of what I will call their "metaphysical aspects"—but don't let the term mislead you. One of the things I aim to show is that none of the considerations I have in mind really go beyond the central concerns of *physics*, of science, as ordinarily understood.

This aspect of scientific theories, their picture or conception of the world, is going to be the focus of this chapter. My primary aim is really just to reintroduce this into philosophical discussions of theoretical equivalence. This does not exactly contradict the literature on formal equivalence, which can be seen as proposing a necessary, if not sufficient, condition on the equivalence of physical theories. It is nonetheless worthwhile explicitly defending the importance of a theory's conception of the world, since recent discussions do not generally mention it, and what they do say can easily leave you with the impression that formal equivalence is intended to be sufficient for wholesale theoretical equivalence. In this way I will be joining a recent minority chorus of philosophers who have been arguing against formal accounts of equivalence, although I do so in a different way, and I will come to some different conclusions.[2]

Despite my going against the grain of recent literature, one conclusion I will be coming to should not, I think, be all that controversial: which is that when it comes to the equivalence of *scientific* theories, we mustn't lose sight of all aspects of these

[1] Formal accounts include the proposals of Quine (1975) and Glymour (1970, 1977, 1980) (see Barrett and Halvorson (2016b) for discussion) as well as more recent ones that extend those ideas to further cases, using more sophisticated mathematics, as in Halvorson (2012, 2016, 2019); Barrett and Halvorson (2016a, 2017); Weatherall (2016a,b, 2017); Teh and Tsementzis (2017); Barrett (2020b); Nguyen et al. (2020). Hudetz (2019) is a recent formal account that aims to be supplemented by a criterion of interpretational equivalence. A related but distinct notion in physics is that of a duality (de Haro, 2017, 2020; de Haro et al., 2017; de Haro and Butterfield, 2018). In physics, dual theories are often said to be "the same theory" even though, as Butterfield (forthcoming) argues, the theories are not interpreted to say the same things about the world, and so are not equivalent in any ordinary sense of the word. A recent survey of various philosophical approaches to equivalence in physics, formal and not, is in Weatherall (2019a,b).

[2] That chorus includes Ruetsche (2011, Chs. 1–2); Coffey (2014); van Fraassen (2014); Nguyen (2017); Bradley (2019); Butterfield (forthcoming); Teitel (forthcoming). See also Sklar (1982).

theories, including those that go beyond the formalism, or even the formalism plus empirical content. However, this will lead me to a further conclusion that is bound to be more controversial: which is that cases of genuine equivalence between physical theories are harder to come by than many people, especially many philosophers of physics as well as physicists themselves, seem to think.[3] That said, even this more controversial conclusion, I aim to show, follows from considerations familiar from ordinary science.

Although the topic of theoretical equivalence dovetails with the issue of the underdetermination of theory by evidence, I will only obliquely address the latter here, in part by contrasting my conclusions with those of Norton (2008b). In one way, my more controversial conclusion is in agreement with Norton's discussion; in another way, it is diametrically opposed to it. Norton argues that there aren't many cases of genuine underdetermination in science—cases of theoretically inequivalent, yet epistemically equivalent scientific theories—on the grounds that there aren't many cases of genuine theoretical inequivalence. He surveys several pairs of theories that are alleged to be instances of underdetermination, and concludes (for reasons we will see) that they are instead cases of theoretical equivalence. There is then no conclusion of underdetermination supported by these cases, for there is no genuine choice to be made: the two theories are fully equivalent, the same theory in different guises.

Against Norton, I will be suggesting that there are a number of cases of genuine theoretical inequivalence—more than most philosophers of science will be comfortable with. Nonetheless, I agree that cases of genuine underdetermination are hard to come by, albeit for a very different reason. The reason is that the inequivalent theories will typically not be epistemically equivalent: there is often (extra-empirical) evidence in favor of one over another, such as the sorts of structural considerations I have been discussing in this book. (What of instances where the inequivalent theories are epistemically equivalent? Those are simply cases of rotten luck. I can see no reason the realist must be able to guarantee our ability to unearth all the physical facts about the world.)

I am not going to try to offer an account of theoretical equivalence in what follows; instead I will rely on a rough, pre-theoretic, intuitive idea that two theories are equivalent when they "say all the same things" but perhaps in different ways. Nor will I take a stand on the different formal accounts that have been proposed. As will be clear from earlier chapters, I do suspect that a formal equivalence of the right kind is a necessary condition on theoretical equivalence, but that won't matter for purposes of this chapter. And although the topic of theoretical

[3] Metaphysicians seem more comfortable with cases of inequivalence. Consider the following from McSweeney: "one of my starting assumptions is that if two theories seem to be saying distinct things about the world, we need a defeater for that seeming in order to get an equivalence claim—our default setting should be inequivalence in most cases" (2016, 274).

equivalence seems to presuppose that we know what scientific theories are, I won't take a stand on that either. Everything I say should be compatible with whatever your preferred account of scientific theories, whether semantic or syntactic or what have you. (The terminology in the literature can make these discussions very confusing. Different people mean different things by "theory," which on its own can result in different conceptions of theoretical equivalence. It will become clear that by "theory" I don't mean the bare formalism alone, but something that includes a conception of how the formal apparatus hooks up to the world.[4]) The idea of scientific explanation is going to play a pivotal role in the discussion too, yet the general points I wish to make hold in the absence of any particular account of scientific explanation, even if the details of those points may vary depending on one's preferred account. I won't take a position on scientific explanation here.

Finally, recall that I am assuming some version of scientific realism. It will quickly become clear that one way of avoiding my conclusions is to adopt an antirealism toward our scientific theories. Since I am a committed realist, I will be happy enough if you think that antirealism is the only way to avoid these conclusions. (At the same time, recall from Chapter 1 that certain antirealists will be able to agree with my conclusions.)

7.2 Initial cases

I am going to proceed by considering a series of cases, roughly in order from least to most controversial. I am going to do this even though it is virtually impossible to find any agreed-upon case that can be used as a starting point in these discussions. As a result, none of what I say will rise to the level of knock-down argument. Yet I hope to convince you that what I say about these cases has a certain level of plausibility, that my take on them is in line with some familiar physical reasoning. That is all I will need.

Start with a case that is reasonably uncontroversial among philosophers. Take the standard textbook or "Copenhagen" version of non-relativistic quantum mechanics and compare it with Bohm's theory. These theories are generally regarded as empirically equivalent, yielding the same probabilistic predictions for experimental outcomes.[5] But hardly anyone (at least in philosophy) would

[4] You might think that formal accounts must be assuming otherwise, equating a physical theory with its formalism. But proponents of formal accounts do not seem to be doing that, and to the extent that they are, so much the worse for their claim to be capturing a notion of equivalence that is relevant to science and scientific theories.

[5] Norton (2008b) says that they are not strictly observationally equivalent, since in Bohm's theory (unlike textbook quantum mechanics) particles always have definite positions and the wavefunction never collapses; as a result, there is an (extremely small) chance of there being an observable effect on particle behaviors of the uncollapsed part of the wavefunction. I won't try to settle on the right notion of empirical or observational equivalence here. A rough idea will suffice for our purposes, something

say that these are fully equivalent theories. According to Bohm's theory, there are particles that always have definite positions and behave deterministically. These things are famously denied by the orthodox theory, which many people think does not even rise to the status of a coherent or sufficiently precise physical theory, but in any case denies that particles are like this. According to orthodoxy, there is often no fact of the matter about a particle's position, except in certain measurement situations, when things are governed by an indeterministic law of wavefunction collapse. Even granting that both theories say that there are particles at the fundamental level, these theories disagree about what these particles are like and how they behave.

As a result of these differences, the two theories give very different explanations of the phenomena. Why do we see a particular distribution of marks on a fluorescent screen at the end of a two-slit experiment with electrons? Bohm's theory says: because electrons always have particular, definite spatial trajectories, which evolve deterministically, in a way determined by the wavefunction; so that given their initial positions and wavefunctions, the electrons will be led to bombard the screen according to that distribution. Orthodoxy says, in effect, that nothing in between the emission of the electrons and the screen explains what we observe. In particular, there are no facts about electrons' locations between the source and the screen that explains why they landed in the way they did. Instead, what we observe is explained by means of what happens indeterministically when we "measure" the particles' positions at the location of the screen.

You might think that there must be a formal or mathematical inequivalence that is responsible for the theoretical inequivalence in this case. Yet I want to emphasize the fact that we seem to come to a reasonable judgment of non-equivalence independently of any formal considerations. Even setting any formal differences aside, there is a clear metaphysical, or just plain *physical*, difference between orthodox and Bohmian quantum mechanics.

Call this sort of difference a *metaphysical inequivalence* between theories: a difference in their pictures of the world—in what there is, what it is like, and how and why it behaves in certain ways to give rise to what we observe. (Notice how close this comes to what one might have thought of as our pre-theoretic notion of theoretical inequivalence, making it all the more surprising that recent discussion has focused instead on theories' formal features.) The suggestion is that a metaphysical inequivalence suffices for theoretical inequivalence. Despite my use of the word "metaphysical" in characterizing the differences between Bohm's theory and orthodoxy, notice, the judgment of inequivalence in this case stems from considerations familiar to ordinary physics, such as a difference in the

like: yielding all the same observable predictions in ordinary experimental situations. The two theories are empirically equivalent in this sense.

theories' laws (there is no fundamental law of wavefunction collapse in Bohm's theory[6]) and their explanations of the phenomena.

This sort of metaphysical inequivalence holds among all the theories of quantum mechanics that philosophers discuss and debate: not only these two, but also other collapse theories, many worlds, many minds, and so on. If a physical theory is an account of what there is in the world, what it is like, and how and why it behaves in certain ways, then none of these are fully equivalent physical theories, because they are not metaphysically equivalent: they disagree about what there is, what it is like, and how and why it behaves in various ways. They are not just different presentations of the same physical theory; they contain genuinely different accounts of the physical world.

As mentioned earlier, no case is completely uncontroversial. Norton (2008b, 37) notes that there are arguments one could give that Bohm's theory contains extra structure compared to orthodoxy; that this structure is superfluous; and that it therefore does not correspond to anything physical, but is just excess mathematical noise—a view that allows us to see the two theories as completely equivalent, mere notational variants. (So long as we do not take the superfluous structure itself to be a marker of inequivalence, contrary to what I have suggested in earlier chapters.) Yet that kind of argument just can't work, because the metaphysical inequivalence suffices to conclude that the theories are not equivalent, independently of any structural considerations. As it happens, in this particular case, most philosophers will agree with this conclusion, which is why I am using this as a reasonably uncontroversial example of theoretical inequivalence in physics.

Turn to another case, which will start to get a little more controversial. Consider classical Newtonian mechanics, and think of the nature of space and time according to this physics. Where Newton thought that his physics requires an absolute space that persists through time, we now know that the theory can be formulated in a different way, in terms of a four-dimensional Galilean spacetime, which does away with Newton's absolute space and corresponding preferred standard of rest.

Now consider two theories of Newtonian mechanics, one set in absolute space and time (alternatively, Newtonian spacetime), the other in Galilean spacetime. And assume substantivalism, that space and time, or spacetime, exists (in the terms of Chapter 5, that spatiotemporal structure is fundamental). These theories are empirically equivalent by any reasonable standard. Newton's laws entail that we could never detect the preferred frame, if there were such a thing. No matter which inertial frame you choose, the laws always predict the same results: Newton's laws are invariant under transformations of inertial frame.

Despite their empirical equivalence, these are not fully equivalent theories, not mere notational variants. Here, too, you may wish to point to a formal

[6] There will however be a nonfundamental law of "effective collapse": Dürr et al. (1992).

inequivalence between them—they seem to disagree about the structure of space and time, for instance—but even setting that aside, the theories are not metaphysically equivalent. They agree on a fundamental particle ontology and the basic laws (though one could question the complete equivalence of the laws, for reasons to come). They nonetheless have different pictures of the world, resulting in different explanations of the phenomena. They disagree about whether the water sloshes up the sides of Newton's bucket because it is rotating with respect to absolute space, for example. More generally, in Newton's own theory the existence of absolute space explains the difference between inertial and non-inertial motion. This is explained in a different way by the Galilean spacetime version of the theory, which will not make any reference to absolute space—there is no such thing. (Instead it will refer to a four-dimensional inertial structure.)

These theories disagree about what the world is like and why physical systems like Newton's bucket behave in the ways we observe them to. The theories point to different things in the world as being responsible for the phenomena. The two versions of Newtonian mechanics are not metaphysically equivalent, in other words, so they are not fully equivalent.

What I said about this case will be more controversial than what I said about orthodox quantum mechanics versus Bohm's theory. Norton argues that the absolute space of Newton's theory is superfluous—we now know that the physics does not really need it—so that it must not correspond to anything physical. (More exactly, he says that each of the different possible preferred standards of rest would be superfluous, but it is a short step from this to the conclusion that absolute space is itself superfluous.) The two theories of Newtonian physics can then be seen as fully equivalent—variant formulations of the same physical theory, alternative presentations of the same physical facts. One of them simply contains excess mathematical noise that doesn't correspond to anything physical. Or at least, "we cannot preclude the possibility" (2008b, 19) that they are completely equivalent, as Norton more mildly puts it.

To say that the extra structure of Newton's theory is merely excess mathematics, however, strikes me as a distortion of Newton's own theory. It is much more natural to say that Newton *was* assuming different physical facts—he thought that there must be absolute space, in order to explain the behavior of the water in the bucket—and that (we now know) he was wrong about what's required in the world in order to explain this. The two versions of Newtonian mechanics are not just different presentations of the same theory, because they are metaphysically (and likely also structurally) inequivalent. At the very least, you should agree that this is *a* plausible take on Newton's own theory, and according to it, the two versions of the theory are not fully equivalent.

What then happens to the threat of underdetermination? In this case, as we saw in Chapter 3, there are good reasons to infer the theory that does away with the additional absolute space structure. The two theories are not epistemically equivalent, in other words. They are not underdetermined by all the evidence, albeit some of that evidence is extra-empirical.[7]

This brings to mind a difficult question, alluded to in Chapter 3. Philosophers of physics generally believe that a formulation of Newton's theory in terms of Newtonian spacetime, rather than absolute space and time as I discussed it above, is a good representation of Newton's *own* theory, the physics as Newton himself saw it. This thought is standard nowadays in philosophy of physics, and so far in this book I have not questioned it. However, given the sorts of considerations coming to light in this chapter, it may be that the Newtonian spacetime formulation is more accurately characterized as a distinct theory from the one in terms of absolute space and time (while capturing much of the same physics), contrary to philosophical consensus. In other words, there may be three distinct theories here: Newtonian physics set in absolute space and time; in Newtonian spacetime; and in Galilean spacetime. I won't address this further here, but wish to point out that there are reasons for such a viewpoint, as will become clearer by the end of the chapter.

Turn to a third case. Consider Einstein's theory of special relativity and compare it with Lorentz's ether theory. Where Einstein's theory does away with absolute simultaneity, Lorentz's theory posits the existence of an ether, whose absolute state of rest corresponds to an absolute simultaneity structure that picks out a preferred simultaneity frame.

These theories are empirically equivalent on any reasonable understanding of the notion. But they are not wholly equivalent. Again, there seems to be a formal or structural inequivalence between them (as discussed in Section 3.3). Yet even setting that aside, there are clear metaphysical, or just plain physical, differences between them. They have very different pictures of the world, which result in different explanations of the phenomena, such as the null result of the Michelson–Morley experiment, and more generally of the fact that we do not detect an absolute simultaneity frame. According to Lorentz's theory, there

[7] It is worth noting that Norton's aim is different from mine, and as a result, he does not come out and assert the equivalence of the theories, but is content to point to the possibility of their being equivalent. Norton wants to undermine various underdetermination arguments about science, by showing that pairs of observationally equivalent theories are typically not examples of genuine underdetermination, on the grounds that we cannot preclude the possibility that the theories are in fact equivalent. Given his stated aim, Norton does not need to argue that the two theories *are* notational variants. All he needs is that we have no good reason to conclude that they are not; or more forcefully, that the two theories are "very strong candidates for being variant formulations of the same theory" (2008b, 35). That said, he comes awfully close to concluding that the pairs of theories he discusses are notational variants. He even suggests that the conclusion of theoretical equivalence is usually the reasonable one to make, because the theories' very observational equivalence typically requires an equivalence of theoretical structure, as he puts it. He furthermore says that, "Many observationally equivalent theories differ on additional structures that plausibly [represent] nothing physical" (2008b, 19), and the "additional structures will be strong candidates for being superfluous, unphysical structures" (2008b, 35). In all, he comes very close to endorsing the claim that the given pairs of theories are notational variants, for all intents and purposes saying just that.

are facts about absolute simultaneity (objective, frame-independent facts about whether two spacelike separated events are really simultaneous with each other), but objects shrink and time dilates in just the right ways, depending on objects' velocities relative to the ether, to prevent us from ever detecting those facts.[8] According to Einstein's theory, we observe a similar kind of shrinking and dilating, but these are frame-dependent effects of objects' velocities relative to one another. In particular, these observations have nothing to do with an ether, and absolute simultaneity is undetectable because there is no such thing: there is nothing in the physical world corresponding to it.

Norton argues that (we cannot preclude the possibility that) these theories are fully equivalent, on the grounds that the extra structure of Lorentz's theory—the preferred simultaneity frame—is superfluous to the physics, as we now know; so we may reasonably conclude that it does not correspond to anything physical in the world. It is merely mathematical excess, and the two theories are in fact notational variants.

As in the case of Newton's physics, however, this strikes me as a distortion of Lorentz's theory. It is much more natural to see the ether theory as distinct, with additional structure that is taken to correspond to something physical—an absolute simultaneity structure corresponding to the state of rest of the ether.[9] At the least, this is *a* plausible take on Lorentz's theory, and according to it, the theory is not wholly equivalent to Einstein's theory. And whereas Norton says that his view aligns with what the physics community has concluded about this case, I think it is equally if not more natural to say that physics has concluded that Einstein was *right* and Lorentz was *wrong*; that they disagreed about the physical facts, and only one of them was getting those facts right. (Norton acknowledges that it is always *possible* to interpret "logically distinct" theories as physically inequivalent. The suggestion is that this is the right thing to do in this case, given a natural understanding of the theories.) Note, too, that as in the case of the different versions of Newtonian physics, the threat of underdetermination is alleviated by the fact that there are good reasons to infer one theory over the other, despite their empirical equivalence.

This goes against the view of Brown (2005) and Albert (2019b, 2020), mentioned in earlier chapters, according to which Einstein's and Lorentz's theories are not only equivalent, but the very same theory. For the theories have the same dynamics, and on their view there is no such thing as a pre-dynamical

[8] As Bell puts it, Lorentz, contra Einstein, "preferred the view that there is indeed a state of *real* rest, defined by the 'aether', even though the laws of physics conspire to prevent us identifying it experimentally" (1987a, 77; original italics).

[9] Bradley (2019) similarly argues that the extra structure of Lorentz's theory "play[s] an essential role in the conceptual framework of the theory." She further argues that Lorentz's theory does not have superfluous structure in the sense that Norton has in mind, not only because of its essential conceptual role, but also because removing the preferred standard of rest of Lorentz's theory does not yield the Minkowski spacetime of Einstein's theory: recall the discussion in Section 3.3.

spacetime structure that could serve to distinguish them: a theory's spacetime structure is just a statement about the dynamics. To my mind, this underplays what significant metaphysical differences there are between the theories, as revealed by their divergent explanations of the phenomena.

Although in each case so far—Bohm's theory versus orthodox quantum mechanics; Newtonian mechanics set in absolute space and time versus Galilean spacetime; Einstein's theory of special relativity versus Lorentz's ether theory—one can argue that there is a formal or structural inequivalence between the theories, I want to reiterate the fact that any judgment of non-equivalence seems to rest at least as much on their metaphysical inequivalence. In each case, the theories' different pictures of the world and divergent explanations of the phenomena—what I have been calling, at the risk of being misleading, the theories' "metaphysical" aspects—makes it hard to see them as merely notational variants of a single theory, just alternative presentations of the same physical facts.

You might continue to insist that the theoretical differences ultimately stem from structural ones—a difference concerning particle positions in the quantum case; absolute space in the Newtonian case; absolute simultaneity in the case of Lorentz versus Einstein—and then go on to argue that this must be true of any case of theoretical inequivalence in physics. I do not have a conclusive argument against such a position, but I do want to emphasize that ordinary science seems to be on my side. Familiar considerations from ordinary science, such as how a theory *explains* what we observe, suggest the existence of metaphysical differences between theories that do not—need not—bottom out in purely formal ones. At the very least, we start out with the feeling that these are genuine differences between theories, and it is not clear that these must always arise from structural or other kinds of formal differences. This will be further underscored by considering some additional cases.

7.3 More cases

Turn now to two examples of pairs of theories that are arguably both formally and empirically equivalent, but are nonetheless still not fully equivalent. This will bring us into more controversial territory.

Consider first the GRW theory of quantum mechanics (so-called after the theory's originators, Ghirardi, Rimini, and Weber (Ghirardi et al., 1986)). This theory is like orthodox quantum mechanics in positing an indeterministic collapse of the wavefunction, but it differs from orthodoxy in that this happens as a matter of fundamental law, with fixed probability per unit time, not because of anything having to do with "measurement" or "observation."

There are different versions of the theory, which agree on the dynamics of the wavefunction, but disagree on what there is in space and time and how it behaves. One version posits a mass density that almost always evolves continuously, but every once in a while undergoes a discontinuous "hit" or "collapse" that causes it to clump up within a very small spatial region. Another version posits what has come to be called a "flash" ontology of discrete events. (A flash occurs whenever a hit to the mass density occurs in the other version of the theory.) The flashes are the only things that exist in space and time, and they are surprisingly sparse: at almost all times, space will be completely empty.[10]

These two theories are generally taken to be empirically equivalent, making "always and exactly the same predictions for the outcomes of experiments" (Allori et al., 2008, 362). (This might sound incredible, but each was devised to yield the same predictions of experimental outcomes that ordinary GRW, without the mass density or flashes, does.) They are also arguably formally equivalent, since they assume the same fundamental dynamics governing the wavefunction. (This will depend on one's particular conception of formal equivalence (as well as some further issues I aim to be neutral about in this chapter[11]), but keep in mind that the kind of formal equivalence in play here is likely fairly abstract or high-level: theories that might seem to suggest non-isomorphic descriptions of the world may nonetheless be formally equivalent since the relevant equivalence applies at a more abstract level—recall the case of Heisenberg versus Schrödinger pictures of quantum mechanics.)

Despite any equivalence in those respects, however, these theories do not seem to be mere notational variants. They have very different pictures of the world, and correspondingly different explanations of the phenomena. According to one theory, the distribution we see on the screen at the end of the two-slit experiment results from a continuously evolving mass density that on each run of the experiment passes through both slits, but undergoes a discontinuous change at the location of the screen that causes it to clump up to a tiny spot; repeated runs of the experiment then yield the pattern we eventually see. According to the other theory, it is extremely likely that nothing at all is happening in between the source and the screen in any run of the experiment, until a flash suddenly occurs at the corresponding spot on the screen. Very different things are happening in the course of this experiment in particular and the history of the world in general. On the flash version of the theory, space is almost always empty! Whereas on the mass

[10] The mass density version is suggested in Benatti et al. (1995). The flash ontology was suggested by Bell (1987b, 205) and a version of the theory worked out by Tumulka (2006).

[11] The wavefunction realist will likely see them as formally equivalent. Proponents of the primitive ontology approach (see Allori et al. (2008); Allori (2013, 2015b)) may disagree, given the different equations for the evolution of the matter density as opposed to the flashes, which on this view corresponds to the real physical ontology of each theory. Even then, this will depend on exactly what the formal differences are, and on whether these amount to a genuine formal inequivalence according to one's preferred account of the notion. (If so, there are other theories that can arguably be used to make the case, such as many minds versus many worlds.) This will also turn on subtle issues about how to understand the relationship between a theory and its relativistic extension, since only the flash version of the theory has been shown to be extendable to a fully relativistic spacetime: discussion in Maudlin (2011, Ch. 10).

density version, there is always a massy gloop smooshing around, which from time to time clumps up.

This suggests that empirical and formal equivalence together do not suffice for wholesale equivalence. It matters equally whether the theories agree on what there is in the world, what it is like, and how and why it behaves in various ways to give rise to what we observe. It matters equally, that is, whether the theories are metaphysically equivalent.[12]

Given the extent of the differences between the mass density and flash theories, you may wonder whether we should even consider them to be varieties of a single type of theory called "GRW quantum mechanics." There are good reasons for the umbrella term, however. These two theories have much more in common with each other than with other theories of quantum mechanics. (And they have much more in common with other theories of quantum mechanics than they do with any classical theory.) There are principles peculiar to the GRW theory, which the flash and mass density theories share, and which other theories of quantum mechanics do not, such as a fundamental indeterministic law of wavefunction collapse. This justifies our considering them instances of one overarching type of theory, even if they are at the same time ultimately distinct physical theories.

This goes for all the cases discussed in this chapter. The pairs of inequivalent theories can be seen as different versions of one umbrella type of theory, sharing certain key principles, while at the same time amounting to distinct theories, given their other significant differences. There is a reason the absolute space and time and Galilean spacetime versions of Newton's physics are each considered a theory of "Newtonian mechanics": they have enough in common, including the fundamental dynamics and particle ontology, that we justifiably regard them as instances of a single overarching type of theory, despite the fact that they are not mere notational variants. Likewise for Einsteinian and Lorentzian electrodynamics, or any of the other cases considered here. (I will not attempt to give an account of the delineation of theories, which I don't in any case think would be particularly interesting or fruitful. As mentioned in Chapter 3, there may not always be a fact of the matter about what the key principles of a given (type of) theory are. Better to focus on the different respects in which theories are, or fail to be, similar to one another.)

Turn to a second case of theories that are arguably both formally and empirically equivalent, but are nonetheless not wholly equivalent. This case concerns different formulations of Newtonian gravitation. According to traditional Newtonian gravi-

[12] The primitive ontology approach (note 11) should deem them inequivalent for the above sorts of reasons. Maudlin says that they "are really quite different physical theories despite their empirical equivalence" (2019, 121). The wavefunction realist might not think that the theories are genuinely distinct on the grounds that the difference lies at the level of nonfundamental ontology, though that will depend on exactly what one thinks is in the fundamental ontology of the theories and on one's view about the relationship between fundamental and nonfundamental ontology.

tation, on the usual understanding, masses interact by means of an attractive force, with a magnitude that is proportional to the product of their masses and inversely proportional to the square of the distance between them, and everything happens against a fixed, flat background spacetime, or flat Euclidean space plus time. (Again, set aside the issue of whether the spacetime and space-and-time versions are two different formulations of the same theory.) There is also a "geometrized" version of the theory, called geometrized Newtonian gravity or Newton-Cartan theory, which, in the manner of general relativity, eliminates any reference to a force of gravity. Instead, spacetime curves in the presence of matter, which in turn affects the behavior of matter.[13]

As mentioned in Chapter 6, Weatherall (2016a) argues that these are fully equivalent theories. They are empirically equivalent, allowing for all the same particle trajectories, and they are structurally equivalent in a way that he demonstrates. (There is an equivalence between the categories of their models.) He concludes that these are different presentations of the same theory, alternative descriptions of the same physical facts. (More exactly, he says this of a certain conception of the traditional theory, while allowing that there is another conception on which the theories are not equivalent by his criteria: below.)

He is not alone in this verdict.[14] Knox (2014), for different reasons, also intimates that these are equivalent theories. She argues that the same reasons leading us to regard Galilean spacetime as the best spacetime setting for traditional Newtonian physics should likewise lead us to regard the curved spacetime of geometrized Newtonian gravitation as the best spacetime setting for the traditional gravitational theory. In other words, traditional Newtonian gravitation "itself is best interpreted as a curved spacetime theory, albeit written in a form that obscures its geometrical structure" (Knox, 2014, 878). Traditional and geometrized Newtonian

[13] Recall Chapter 1, note 4. On the geometrized theory, the gravitational force is effectively incorporated into the affine connection of the spacetime: the connection, which depends on the distribution of matter, is allowed to be nonflat, and freely falling particles follow the geodesics of the connection. I set aside the issue of potential inconsistencies in the traditional theory when applied to homogeneous cosmologies (Norton, 1993b, 1995; Malament, 1995), even though this is not entirely irrelevant to the current discussion. Malament argues that the inconsistency is only apparent since it is based on a particular formulation of the theory, and disappears when we move to the geometrized formulation—an argument that seems to presuppose that they are notational variants. However, in the end Malament says that it does not really matter whether we consider them to be equivalent formulations or genuinely distinct theories, for his primary conclusion holds either way (namely, that from the vantage point of the geometrized formulation, we can see that the alleged inconsistency is just an artifact of formulation).

[14] There is surprisingly wide disagreement on this case. Glymour (1977), Jones (1991), and Earman (1993) take them to be inequivalent. Malament (2007) and Jeffrey Barrett (2008) effectively treat them as equivalent. (Malament (1995) does not take a firm stance on the question: note 13.) Coffey (2014) argues that there is no fact of the matter about their equivalence, since it depends on how we interpret the theories, and in his view there is no fact of the matter about how to do this. I agree that whether the theories are equivalent will depend on their metaphysical aspects, which are not entailed by the formalism. But I also think there are more or less natural understandings of theories, relative to which we can draw reasonable conclusions about their equivalence.

gravitation are essentially equivalent theories; in particular, they posit the same (curved) spacetime structure. (In this latter respect she departs from Weatherall, who claims there is no real difference between describing the spacetime structure of either theory as potentially curved or not. Knox's discussion is not targeted directly at the question of theoretical equivalence,[15] and she is ultimately non-committal as to whether the geometrized theory counts as a reformulation of the traditional theory or instead a distinct (and improved) theory. However, she does say that it is "natural," given her arguments, to see these "as reformulations of the same theory" (2014, 877), so that her discussion can be read in support of the conclusion that they are equivalent theories.)

Norton, too, suggests that the two versions of Newtonian gravitation are plausibly fully equivalent, on the grounds that the "background inertial structure [of the traditional theory] is physically superfluous" (2008b, 37). Therefore, just as he suggests of previous cases, this structure should not be seen as corresponding to anything physical, but as merely excess mathematics, an artifact of the formalism, and the two versions of Newtonian gravitation are really notational variants. He says that this is in fact the standard view of the matter.

It strikes me as much more natural to see these as distinct theories, however, with different pictures of the physical world—more natural to say that they posit different physical ontologies, since they disagree on the existence of certain forces; a different type of spacetime, since in only one of them can spacetime change and be curved; even different laws, stated in terms of a different spacetime and ontology. As Malament says (but see notes 13 and 14),

> In the geometrized formulation of the theory, gravitation is no longer conceived as a fundamental 'force' in the world, but rather as a manifestation of spacetime curvature (just as in relativity theory). Rather than thinking of point particles as being deflected from their natural straight (i.e. geodesic) trajectories, one thinks of them as traversing geodesics in curved spacetime. (Malament, 2007, 266)

The traditional and geometrized theories contain different pictures of the world, resulting in different explanations of the phenomena. The same type of observed particle motion will in one theory be explained by means of a gravitational force exerted on it by another particle located at some distance, in the other theory by means of the local curvature of spacetime. This doesn't seem like simply two different ways of saying the very same thing.

Notice how this judgment rests on considerations we take seriously all the time in physics. Whether a theory is local—whether one thing can affect another that is at a distance only by means of a continuous causal chain in between—is

[15] Nor is she likely to endorse a formal account of equivalence: see Knox (2011).

usually regarded as a physically significant, distinguishing mark of a theory. The geometrized theory is local in this sense; traditional Newtonian gravitation is not.

Notice, too, that the opposing view—either that describing a world with one type of spacetime structure can be fully equivalent to describing it with an apparently different type of structure plus certain forces, or that Newtonian gravitation is itself a curved-spacetime theory—would eradicate one of the main innovations we often cite for general relativity over Newtonian gravitation, which is that it eliminates the gravitational force in favor of spacetime curvature. When trying to convey what these theories say about the world, especially the differences in what they say about the world, we often say things like: according to general relativity, Newton's apple wasn't pulled to earth by the force of gravity; instead, it was following an inertial, force-free trajectory through the spacetime structure near the surface of the earth. We don't ordinarily take this to be just a redescription of the same physical facts, but a different physical theory, with a different explanation, appealing to different physical facts. This seems to be the usual conclusion in physics anyway.[16]

When it comes to traditional Newtonian gravitation versus Einstein's theory of general relativity, there are of course empirical differences that will lead everyone to regard them as inequivalent theories. (Likewise for geometrized Newtonian gravitation versus general relativity, the former yielding the same empirical predictions that traditional Newtonian gravitation does.) The point remains that ordinary science recognizes the above sorts of explanatory differences between them as well. If we wish to preserve those differences between general relativity and Newton's theory of gravitation, then we should do the same thing for traditional Newtonian gravitation vis-à-vis geometrized Newtonian gravitation: we should take what appear to be differences in the nature and structure of spacetime and in the ontology of forces to be genuine differences between theories. This is not to deny that it is illuminating to learn that something like Newton's gravitational theory can be stated in a way that preserves much of the original theory, while being more like general relativity than we may have thought any theory similar to Newton's could be. (One thing we learn is that general covariance per se is not a distinctive feature of general relativity.) We can recognize this without at the same time obliterating what significant differences there are between the theories, differences that are important to understanding exactly what each theory is saying about the behavior of gravitational systems in the world.

At this point, it is important to be reminded that a theory's metaphysics is not entailed by its formalism. Because of this, there will always be room for disagreement, about this case or any other. We *can* interpret the traditional theory

[16] This kind of thing can be found in textbooks such as Hartle (2003) and Carroll (2004). There is an alternative formulation of general relativity that some construe as a force-based theory: discussion and references in Knox (2011), who ultimately disputes that take on it.

in such a way that it is completely equivalent to a geometrized version. However, on at least *a* natural understanding of the two theories, including a conception of the traditional theory that arguably comes closest to what Newton himself had in mind (arguably, since Newton famously withheld from hypothesizing about the ultimate nature of gravity; see also note 35), they are not fully equivalent, and because they are not metaphysically equivalent. They make substantially different claims about the world, resulting in different accounts of the phenomena. (Recall from Chapter 4 that it is an assumption of the traditional theory, on a natural understanding, that there is a force of gravity, a physical posit on which both general relativity and geometrized Newtonian gravity differ. It would not be unreasonable to identify as key principles of the traditional theory the assumptions that space and time are fixed and flat, and that there is a physically real gravitational force, principles with which the geometrized theory disagrees, even while allowing that the two have enough in common to both qualify as theories of "Newtonian gravitation"; whereas general relativity, which does not share those key principles, does not so qualify.[17])

Jones, for one, agrees that the geometrized and traditional theories of Newtonian gravitation "in some sense 'save the same phenomena,' but with very different explanatory frameworks, that is, very different ontological commitments" (1991, 190). However, he presents this as a case of "ontological ambiguity" in physics that is problematic for the realist, just as we saw in Chapter 4 he suggests of different formulations of classical mechanics. As in the case of different formulations of classical mechanics, the different formulations of Newtonian gravitation are generally taken to be mere notational variants, but with different ontological pictures if interpreted realistically. And yet, Jones suggests, there seems to be no reason to choose one over the other, and so to adopt one ontological picture or the other.

As in the case of classical mechanics, here, too, I disagree that this must spell trouble for the realist, for the metaphysically inequivalent theories need not be epistemically equivalent. There can be good reasons to infer one over the other, as we saw for the particular case of Newtonian versus Lagrangian mechanics, and may be the case for geometrized as opposed to traditional Newtonian gravitation as well, although I won't investigate this here. In any case, as I have said before, it is open to the realist to allow for cases of inequivalent theories that are underdetermined by the evidence. Such cases of epistemic bad luck are unfortunate, but they needn't doom realism altogether.

[17] Compare Glymour on the idea that these sorts of initial posits are essential to a theory: "there are built into our theories various principles that establish presumptions as to the forces acting in various situations, for example, in Newtonian theory that the only significant force determining the trajectories of the major bodies of the solar system is gravity. Such presuppositions are not, and perhaps cannot be, laws, but they are an essential part of our theories nonetheless" (1980, 350).

As it happens, the case of Newtonian gravitation is even more subtle than I have let on. The reason is that there are additional formulations of the traditional theory available. The theory can be stated as above, in terms of a gravitational force that acts directly and instantaneously between distant masses. It can also be formulated differently, in terms of fields. According to a field-based version of the theory, the gravitational interaction is local, mediated by a continuous field that exists in the space in between any two masses, and which (unlike the gravitational force) is present even in the absence of one of the masses. Although generally treated as equivalent by physics texts and by philosophers, the above considerations suggest that, on a natural understanding, these are not fully equivalent physical theories. They posit different things in the physical world as being responsible for gravitational phenomena.[18] They give different answers to a question posed by James Clerk Maxwell:

> The question is that of the transmission of force. We see that two bodies at a distance from each other exert a mutual influence on each other's motion. Does this mutual action depend on the existence of some third thing, some medium of communication, occupying the space between the bodies, or do the bodies act on each other immediately, without the intervention of anything else?
>
> (Maxwell, 1890a, 311)

The force- and field-based formulations disagree on the answer to this question, which gets to the heart of their differing accounts of the phenomena. They are not metaphysically equivalent; so they are not fully equivalent.[19]

There are even different kinds of fields that can be used to formulate a field-based version of the theory: the (scalar) potential field or the (vector) gravitational field. On a natural understanding, all three of these are inequivalent physical theories. (The two field-based theories are closer to each other than to a force-based theory; they answer Maxwell's question in the same way, and are better candidates for wholesale equivalence. That said, the fields they posit, on a natural understanding, possess different natures.) These theories posit different physical ontologies, different physical things that exist over and above massive particles—whether an inter-particle force or a scalar or a vector field—which figure in different fundamental laws: the inverse-square law of the force-based theory, Gauss' law for gravity for the gravitational-field theory, Poisson's equation for the

[18] Stein (1970b) argues that Newton himself was in fact committed to the existence of fields. Mundy (1989, n. 5) argues that Stein has only demonstrated Newton's commitment to fields in a secondary sense, as things definable in terms of the fundamental ontology.

[19] Feynman notes that the usual way of stating Newton's gravitational law has "an unlocal quality. The force on one object depends on where another one is some distance away," whereas the "field way" of stating the law "says a completely different thing" (1965, 50–1). He nonetheless goes on to say that they are "equivalent scientifically" (1965, 53). It is clear from his discussion that he means that they are equivalent in a sense to come in Section 7.4.

potential-field theory. (They each posit a different physical ontology and also a different "nomology," in a term from Maudlin (2018).)

The three formulations have a lot in common. Their laws and fundamental quantities are mathematically inter-derivable. There is a reason we consider each of them a theory of "traditional Newtonian gravitation." Nonetheless, on a natural understanding, they are not completely equivalent, for they are not metaphysically equivalent.[20]

This idea of a natural understanding of theories that I have been appealing to is not intended to be simply the crude or naive method of face value interpretation rejected in Chapter 6. (Though even if it were, the point remains that there are natural enough understandings on which these qualify as inequivalent theories.) Rather, it amounts to just the kind of "reasonably straightforward" method I endorse. We will still be careful to mind the gap between the mathematical formalism and the physical world; to not unthinkingly read everything about the physical world directly from the mathematical formalism; to presume that the formulation on the basis of which we form a natural understanding has been chosen for good empirical reason; and so on. By calling it a "natural understanding," I mean to convey that it is a reasonably clear and straightforward one, not that we are to completely ignore the additional subtleties and complications discussed in Chapter 6. (Compare Maudlin's (2018) discussion of "ontologically clear" theories of classical electromagnetism (in terms of what he calls a "canonical presentation") even though, as we saw in Chapter 6, Maudlin himself rejects the "face value" method of interpretation.)

Since the Newtonian gravitational field does not have a dynamics of its own (by contrast to Maxwell's equations for electromagnetic fields, for example, which further allow for source-free fields), and since Newtonian gravitational interactions occur instantaneously at a distance, among other reasons, many people will conclude that the gravitational field is not what's physically real on this theory; it is just a way of keeping track of the (direct, unmediated) inter-particle forces.[21] (Though the direct action-at-a-distance nature of the forces may lead others in the opposite direction, toward taking the field ontologically seriously.) The gauge freedom in the potentials—the fact that different gravitational potentials seem equally capable of describing the same physical situation—will further suggest to many that this is the least preferable of the three. (By the same token, there could be a gravitational Aharonov–Bohm effect suggestive of the physical reality

[20] Although standardly treated as equivalent, it is also often noted that the law of the potential-field formulation is more general than that of the force-based formulation (as does Malament, 1995, n. 6 and Sec. 5). This is a kind of inequivalence between the two, which could even amount to an empirical inequivalence, depending on how one construes the notion.

[21] Thus Lazarovici: "While it can be a useful mathematical tool, the gravitational field is just a book-keeper of direct particle interactions rather than a physical entity that exists over and above the particles" (2018, 156). This is a reasonably standard view, but an alternative conception of the theory is available, even if there are reasons to think it is a worse theory.

of the gravitational potential in the quantum realm, analogous to the effect that suggests to many the physical reality of the electromagnetic potentials in the quantum realm.[22]) Some will even go so far as to argue that fields in general are not physically real, on any reasonable physical theory, as does Dustin Lazarovici (2018).[23] All of this only serves to underscore the point that these are distinct pictures of the physical world, described by distinct physical theories, each with a different physical ontology.

On a natural understanding, then, the force-based theory, the two field-based theories, and the geometrized theory are all distinct theories of Newtonian gravitation; for none of them are metaphysically equivalent. They give rise to "radically different" accounts of the phenomena, as Jones describes it:

> In the first approach . . . , Newton's law of universal gravitation is usually taken to describe the properties of a fundamental gravitational force which has about it a renowned kind of dual nonlocality. . . . The second approach described characterizes the gravitational interaction in purely local terms, but it does so by means of the introduction of a new physical entity into the explanatory picture—the potential field. This field is eliminated if physical space—heretofore treated (implicitly) as Euclidean—is assumed to be curved by the presence of matter . . . . But then a kind of causal efficacy is associated with the structure of space itself.
>
> (Jones, 1991, 189–90)

(Jones does not distinguish between a gravitational-field and potential-field version of the theory, though his own arguments suggest that he could.)

We can now see a way in which Weatherall's (2016a) conclusion is less radical than it initially seemed. Weatherall argues for the equivalence of traditional and geometrized Newtonian gravitation, assuming both that the traditional theory is formulated in terms of the gravitational potential, and that the theory does not regard differences in the potentials as genuine differences: models that differ only with regard to the gravitational potential are taken to be physically equivalent. On the usual way of understanding such ("gauge") quantities, however, this amounts to denying the physical reality of the gravitational potential.[24] This makes it less surprising that the theory is equivalent in significant ways to a theory that eschews any reference to a gravitational potential. It also makes one wonder whether the equivalence claim is based on the most natural conception of the traditional theory,

[22] See Hohensee et al. (2012). Note that there is also a gauge freedom in the gravitational field (see the references in note 13).

[23] Lazarovici draws this conclusion by focusing on the case of classical electromagnetism, which will be discussed in Section 7.5.

[24] Recall the discussion in Chapter 2 suggesting that this usual understanding is a bit too quick, for quantities that are defined relative to an arbitrary choice (of reference frame, for example) need not be completely unreal or unphysical. Discussion in this chapter brings out a further reason for this: it depends on what we postulate to be physically real; more below.

since one might have thought it a defining feature of the theory that it posits the physical reality of the gravitational potential or field or force. (Weatherall notes that physicists have generally regarded the traditional theory that way.)

What is more, Weatherall acknowledges that a traditional theory that does treat different gravitational potentials as genuinely different will not be equivalent to the geometrized one, by his criteria. He concludes from this that whether traditional and geometrized Newtonian gravitation are equivalent will depend on which models of Newtonian gravitation one takes to be physically equivalent. But this is just to acknowledge that formalism plus empirical content on their own do not suffice to indicate which features are "mere gauge" and so which theories are equivalent. It is to acknowledge that the equivalence of theories depends on what I have been calling their metaphysical aspects.

There are yet further distinctions that could be drawn, for example, between a space-and-time and spacetime formulation of Newtonian gravitation, or between formulations in terms of point-masses versus a continuous mass distribution, and more besides.[25] I won't examine each and every possible such distinction here, but I will note that, given the above considerations, it is at least not obvious that any of these are completely equivalent physical theories.

7.4 Metaphysical and informational equivalence

Before turning to a few more cases, let me pause to offer a diagnosis of what has gone wrong with the focus on formal criteria.

Formal accounts maintain that physical theories are equivalent when there is a formal and an empirical equivalence between them. (The debate then centers on what is the right notion of formal equivalence.) The motivating thought behind these accounts, taking a cue from Quine (1975) and Glymour (1970, 1977, 1980), is that two theories are equivalent when they yield all the same empirical predictions and are inter-translatable: when everything that is said by one can be said by the other, and vice versa, given a suitable translation, just as we might say of a theory written in English as opposed to French. Or in the model-theoretic terms in which this is also often put: when any model of one theory can be suitably transformed into a model of the other, and vice versa.

It is not without reason that such accounts have been defended by philosophers of physics. The basic thought is intuitive. If two theories give rise to all the same empirical predictions, and if we can translate the claims of one theory into those of another, and back, without gaining or losing any information—if we can recover all the same information from each one—then it seems as though the

[25] Compare Maudlin (2018, 8) on a mass density versus point-particle version of classical electromagnetism, with the passage from the former to the latter yielding a "new theory."

theories must have the same content. They must be saying the same things in different ways, "encod[ing] precisely the same physical facts about the world, in somewhat different languages" (Rosenstock et al., 2015, 315–6).[26] Something like this seems implicit in textbook discussions of the equivalence of Heisenberg and Schrödinger formulations of quantum mechanics, for example, or the Newtonian, Lagrangian, and Hamiltonian formulations of classical mechanics, both generally regarded as paradigmatic examples of theoretical equivalence by physicists as well as philosophers.[27]

Formal accounts thus aim to capture a kind of informational equivalence between theories, and they take this to be wholesale theoretical equivalence. Though intuitive to a point, the approach strikes me as misguided. The reason is simply that two physical theories can have the same content in the sense that we can extract or recover the same information from each one, and yet the theories can still be saying different things about the world. Let me explain.

I use the phrase "informational equivalence" deliberately to allude to an idea of Maudlin's (2007a). In discussing the Einstein, Podolsky, and Rosen paper, "Can Quantum-Mechanical Description of Physical Reality be Considered Complete?" (1935), Maudlin points out that there are two senses of "complete" one might have in mind when trying to answer the question in that paper's title. A theory or description can be what he calls *informationally complete*, which means that "every physical fact . . . can be recovered" from it (2007a, 3151). A distinct notion is that of being *ontologically complete*, which means that the description "provide[s]—in a relatively transparent way—an exact representation of all of the physical entities and states that exist" (2007a, 3154).

Importantly, informational completeness and ontological completeness can come apart. Maudlin argues in particular that a quantum theory that does not posit any fundamental physical ontology in ordinary spacetime can be informationally complete, but will not be ontologically complete. Although such a theory can be used to recover all the quantum mechanical facts, in his view it will not directly represent what's physically real, which must be things in four-dimensional spacetime. He is arguing against wavefunction realists, who say that the fundamental physical ontology is instead directly represented by the wavefunction, which is defined on an extremely high-dimensional space.

Maudlin points out more generally that just because a description is informationally complete does not mean that it is ontologically complete. In a

[26] Rosenstock et al. (2015) are talking about two formulations of general relativity in particular, which I turn to at the end.

[27] Examples in physics include Symon (1971) for classical mechanics (see also the references in Section 4.5: textbooks asserting the equivalence of different formulations of classical mechanics generally rely on something like this idea), or Merzbacher (1998, Sec. 14.2) and Messiah (2014, Secs. 8.9–8.10) for quantum mechanics. Compare the idea of "scientifically equivalent" mentioned by Feynman (1965): see note 19.

deterministic world, for example, a description of the initial state plus the laws is informationally complete: from this we can recover all the physical facts about the world, at all times. But such a description will not be ontologically complete. Despite the informational completeness, we do not conclude that all there is in the world is whatever is contained in the initial state.[28]

The lesson for us is the following. There is a difference between a theory's being such that all the physical facts *can be recovered* from it, and being such that those facts are reasonably *directly represented* by it. In a slogan, there is a difference between what a theory *can* say, or what it can be made to say given suitable definitions, and what it *does* say. Or to put what we will see is another spin on the same idea, the fact that something is mathematically definable using a theory's formal apparatus does not entail that it represents something physically real (as an example at the beginning of the next section will make particularly clear).

Putting this in terms of equivalence: two theories can be *informationally equivalent* without being *metaphysically equivalent*—as is the case for Maudlin's examples of a quantum theory with or without a fundamental four-dimensional ontology, and a deterministic theory with or without states other than the initial one. It is also the case for traditional and geometrized Newtonian gravitation. Even though we can recover all the same information or physical facts from each one—they are informationally equivalent, inter-translatable in this way—they are not metaphysically equivalent. In *one* way, the two versions of Newtonian gravitation are notational variants, in that each can be used to recover the same facts: they contain the same information, coded up in different ways. But in another, physically significant way, they are not *mere* notational variants: they present different pictures of the physical world. There is *both* a sense in which they contain all the same physics, *and* a sense in which they differ with respect to the physics—just as there is both a sense in which the description of a deterministic world in terms of the initial state and the laws is complete, and a sense in which it is not; and, in Maudlin's view, a sense in which the wavefunction description is complete and a sense in which it is not.

The problem with formal accounts of theoretical equivalence, then, is the focus on informational equivalence at the expense of theories' metaphysical equivalence, when these are equally important to our judgments of equivalence in physics. We have seen this in the cases discussed so far, and will see it in further examples in the next section. (Whether all the pairs of theories considered in this chapter count as informationally equivalent despite their metaphysical inequivalence will depend

[28] There are further questions that can be raised, for instance about exactly what is the sense of "transparent" (in an ontologically complete description) such that the mathematical entailment of states at other times, given the initial state and the laws, does not count as a transparent description. I merely want the initial intuitive idea in order to motivate the distinction in the following paragraph. Analogous questions can be raised about that distinction too, but the distinction will be clear enough for my purposes, as illustrated in particular by the examples in the next section.

on exactly what one takes to be the informational content of a theory. This is not completely precise, but it is clear enough for our purposes.)

Two final notes before turning to those further cases. The first is to mark an important distinction. In the course of discussing the possibility of a flash ontology for GRW, Bell notes that the centers of the GRW jumps will be "the mathematical counterparts in the theory to real events at definite places and times in the real world (as distinct from the many purely mathematical constructions that occur in the working out of physical theories . . .)" (1987b, 205). Notice Bell's distinction between the "purely mathematical constructions" of a physical theory as opposed to the mathematical representations of things the theory takes to be "in the real world." One way of putting the concern about formal accounts of equivalence is to say that they fail to be sufficiently attuned to this distinction. In emphasizing theories' formal inter-definability, these accounts can be blind to differences between theories that have to do with which formal aspects are taken to represent real things in the physical world. We will see this by example in Section 7.5.

Second, we have come across realist views that won't be moved by these considerations. According to Wallace and Timpson (2010) and Wallace (2012), recall, the mathematical structure of a theory is the sole guide to the ontology or metaphysics more broadly. They suggest that superficially very different formulations of a theory, such as configuration-space and spacetime formulations of quantum mechanics, can be wholly equivalent, ascribing the very same structure and ontology to the world. Such a view does not allow for a notion of metaphysical equivalence between theories that can come apart from their formal or mathematical equivalence. Moreover, to argue that there is a metaphysical difference between the above pairs of theories is to beg the question against the view, for it assumes that there are meaningful questions about metaphysics and explanation that come apart from theories' formal aspects.

Ontic structural realists, mentioned in Chapter 1, will likely also be unconvinced. So long as two theories have the same structure, in some sense—perhaps, as in Worrall's example of Fresnel's versus Maxwell's theory of light, the same form of equations—there will be no further metaphysical respect in which the theories could be said to differ. There is no "picture of the world" according to a physical theory that extends beyond the theory's structure. (The epistemic structural realist may allow that there is such a thing, but presumably thinks that we at least cannot know about any metaphysical/explanatory aspects that go beyond a theory's structure.)

I take the above cases to reveal that ordinary science, at least, is on my side. Ordinary scientific standards tell us both that the above kinds of explanatory differences are genuine differences between theories, and what differences count as explanatory ones—differences that strike us independently of any considerations having to do with theories' mathematical structures. Ordinary science regards differences in how two theories explain the phenomena as genuine differences, and it tends to see these explanatory differences as arising from underlying metaphysical—or, again, just plain physical—differences. The examples in this chapter suggest that this is plausible as an empirical claim about ordinary science and scientific understanding. They also suggest that this is plausible independently of any particular account of scientific explanation.

7.5 Additional cases

With the distinction between informational and metaphysical equivalence in mind, let's turn to some more cases, which will get increasingly controversial. That said, the conclusions I come to will be based on considerations familiar from ordinary physics.

First, take classical electromagnetism. The equations of the theory, Maxwell's equations and the Lorentz force law, can be formulated in different ways: in terms of the electric and magnetic fields, or the scalar and vector potentials. (Or in the spacetime terms I set aside here, in terms of the electromagnetic field or the electromagnetic four-potential.) The two versions of the theory are empirically equivalent. (In the classical domain, that is, setting aside the Aharonov–Bohm effect, which to many is indicative of the physical reality of the potentials in the quantum realm.) The fields and potentials formulations can also be shown to be formally equivalent in various ways. As a result, these are generally regarded as completely equivalent, mere notational variants—the same theory expressed in different ways. This is what is said in physics books.

However, it is also generally thought that only the fields are what's physically real. The potentials are seen as mathematical constructs used for convenience, not corresponding to things in the physical ontology. (The reason is the gauge freedom in the potentials: different potentials, corresponding to the same fields, are equally capable of characterizing the phenomena.) We use the fields, not the potentials, to explain the observable motions of iron filings on a piece of paper near a magnet, to use an example that Maxwell himself discusses. Maxwell describes how the iron filings will align themselves with the magnetic lines of force, suggesting the presence of a magnetic field in the space surrounding the magnet. There is no mention of the potentials in the explanation.[29] (This is the case even though, as

[29] Here is what Maxwell says (note the focus on explanation, though note as well that he focuses on fields versus forces acting at a distance, rather than potentials): "Thus if we strew iron filings on paper near a magnet, each filing will be magnetized by induction, and the consecutive filings will unite by their opposite poles, so as to form fibres, and these fibres will *indicate* the direction of the lines of force. The beautiful illustration of the presence of magnetic force afforded by this experiment, naturally

mentioned in Chapter 1, Maxwell himself formulated the equations using the scalar and vector potentials. It was Heaviside who reformulated the equations in terms of the electric and magnetic fields.)

In Maudlin's terms, the potentials formulation is informationally complete, but only the fields formulation is ontologically complete as well. And it is the ontologically complete description that we refer to in explaining electromagnetic phenomena. In my terms: the potentials and fields formulations of classical electromagnetism are informationally equivalent, but they are not metaphysically equivalent. For that reason, they are not wholly equivalent. (To put it another way, the mathematical fact that the equations can be formulated in terms of the potentials does not on its own mean that physical reality consists of the potentials. More on this way of putting it in Section 7.6.)

Although we are used to thinking of classical electromagnetism as positing a physical ontology of fields, we can imagine taking the potentials to be physically real instead—as we might want to do in the quantum case, given the Aharonov–Bohm effect (as Aharonov and Bohm (1959) themselves suggested), but could do even in the classical case. Of course, there are reasons for not taking the potentials ontologically seriously in the classical theory. Above all, there is the above-mentioned gauge freedom in the potentials. In response, one could try to "fix a gauge" in order to pick out which of the different potentials descriptions is correct, thereby avoiding the indeterminism that would otherwise result from taking the potentials to be physically real, although in that case one might worry that there isn't a sufficiently well-motivated choice of gauge.[30] (An alternative is to be a quotienter about the potentials, in Sider's sense.) Heaviside said of the potentials that it is "best to murder the whole lot" (1892, 482) since they obfuscate what is really going on physically, and in so doing he was able to recast Maxwell's equations in the improved and streamlined version we know today.[31] In all, there are good theoretical reasons to refrain from positing the potentials as physically

tends to make us think of the lines of force as something real, and as indicating something more than the mere resultant of two forces, whose seat of action is at a distance, and which do not exist there at all until a magnet is placed in that part of the field. We are dissatisfied with the explanation founded on the hypothesis of attractive and repellent forces directed towards the magnetic poles, even though we may have satisfied ourselves that the phenomenon is in strict accordance with that hypothesis, and we cannot help thinking that in every place where we find these lines of force, some physical state or action must exist in sufficient energy to produce the actual phenomena" (Maxwell, 1890c, 451–2; original italics).

[30] Discussion in Belot (1998). Maudlin (2018) discusses the theoretical ramifications of different choices of gauge, which could provide grounds for choosing one.

[31] In the preface to his (1893), Heaviside discusses the fact that he reformulates the theory in terms of electric and magnetic fields and forces "instead of the potential functions which are such powerful aids to obscuring and complicating the subject, and hiding from view useful and sometimes important relations." More thoughts along these lines can be found scattered throughout Heaviside (1892, e.g. 173, 481–5, 511), where he emphasizes that thinking clearly about what the theory is saying physically is what led him to the better formulation of the equations. This is an illustration of how taking what I have been calling theories' metaphysical aspects seriously can have far-reaching consequences for ordinary physics. Thanks to Marc Lange for the pointer to Heaviside's writings on this.

real. The point remains that the potentials and fields formulations of classical electromagnetism are distinct physical theories, with different physical posits. They contain different pictures of the physical world lying behind electromagnetic phenomena.

Here, too, the distinction from Section 7.4 allows us to pinpoint a sense in which the two formulations are not completely equivalent, even while granting that they are equivalent in the respects that lie behind the standard view—which is to say that they are informationally equivalent but not metaphysically equivalent; in particular, only one of them is metaphysically accurate. There is *both* a sense in which they describe the same physics, and a sense in which they differ with respect to the physics. They are notational variants in that we *can recover* the same physical facts from each one; but at the same time, they are not *mere* notational variants, in that they depict different physical realities. The two formulations have a lot in common; there is a reason we consider each one a "theory of classical electromagnetism." They are nonetheless not completely equivalent. They differ in ways that matter to the physics and its accounts of electromagnetic phenomena.

Weatherall (2016a, Secs. 4–5) argues that the potentials and fields formulations of classical electromagnetism are completely theoretically equivalent. However, what Weatherall shows is that a formulation given in terms of the Faraday tensor, which represents the electromagnetic field in spacetime, is equivalent, by his criteria, to a formulation in terms of the four-vector potential, where the latter, as he presents it, treats the vector potential as a gauge quantity: different vector potentials corresponding to the same Faraday tensor are not regarded as physically distinct. What he shows, in other words, is that a fields formulation of classical electromagnetism is equivalent, by his lights, to a formulation that is naturally understood as denying the physical reality of the potentials. As in the case of Newtonian gravitation, this makes the equivalence claim neither very surprising nor counter to the arguments here. For all that Weatherall has shown, formulations that do disagree on the physical reality of the potentials may well be inequivalent.

Kevin Coffey (2014) says that classical electromagnetism provides an example of what he calls "asymmetrical equivalence." Although the two formulations of classical electromagnetism are fully equivalent, it is also the case that only the fields formulation is basic or primary. The potentials formulation is a *re*formulation of the fields one, and not vice versa, since only the fields are physically real. (He notes that this is puzzling, and searches for an account of equivalence that can explain it.) I think we should say this instead. Equivalence is (by definition!) a symmetric relation. It is just that there are different respects in which things can be, or fail to be, equivalent. What Coffey sees as a puzzling case of asymmetrical equivalence is instead a case of equivalence in certain respects and inequivalence in others; in particular, a case of informational equivalence but not metaphysical equivalence.

Nor are the fields and the potentials the only two reasonable physical ontologies for classical electromagnetism. Brent Mundy (1989) argues that the standard formalism can be just as easily construed as a theory of electromagnetic (tensor) forces, which act directly between particles at a distance (retarded action at a distance in particular[32]), rather than a theory of fields or potentials at all. Lazarovici (2018) likewise advocates a force-based, direct interaction conception of the theory (although he endorses the time-symmetric, half-retarded/half-advanced Wheeler–Feynman absorber theory rather than Mundy's retarded-action theory).[33] Whatever one thinks of its merits, a force-based theory is a *distinct* theory, even though it gives rise to the same observable particle motions, and even if it is formally equivalent to both the fields and potentials formulations. It posits a different physical ontology and different fundamental laws—in Mundy's theory, for example, there is a fundamental "distant action force law" analogous to Newton's inverse square law for gravitation—and different explanations of the phenomena (electromagnetic interactions are not spatially and temporally local, for instance).

Here is a different reaction you might have to the case of classical electromagnetism. Set aside distant action theories and focus on a formulation in terms of fields as opposed to one in terms of potentials. You might say that these formulations are *both* informationally and metaphysically equivalent, on the grounds that we have *stipulated* from the beginning that the theory is fundamentally about the fields. Given this initial physical posit or stipulation, the potentials are clearly just alternative mathematical ways of saying things about the fields. (Calling such a thing an initial posit or stipulation is not meant to suggest that it is made in the absence of any empirical or other scientific considerations, as though we can conjure up a reasonable scientific theory entirely by initial a priori stipulation, but to convey that it does not strictly follow from such considerations. Recall the point from Chapter 6 that the physical ontology and the mathematical formulation of the laws (itself chosen for good scientific reason) constrain each other.) The two formulations are then fully equivalent, after all.

Again, there is no algorithm that takes us from a theory's formalism to its metaphysics. We *can* interpret the potentials formulation in this way—as simply another, roundabout way of saying things about the fields. Still, there is *a* natural understanding on which it posits the potentials in the physical ontology. (That there is such an understanding available is underscored by the Aharonov–Bohm effect, which might lead one to the analogous posit in the quantum case.) And on this understanding, the potentials formulation is a distinct physical theory from the fields formulation, with a different picture of physical reality. (In other words, the two ways of understanding the potentials formulation, as either an indirect representation of the fields or as a direct representation of physically real

[32] Mundy discusses the theory set in Minkowski spacetime, so that the action occurs along the light cone, hence not instantaneously.

[33] See Maudlin (2018) for further variations of the theory.

potentials, are in fact distinct, inequivalent theories.) So the point remains that informational equivalence can come apart from metaphysical equivalence, and that these are both important to our judgments about equivalence in physics.

A further subtlety. Above I allowed that if we start out assuming that physical reality contains fields, not potentials, then the potentials and fields formulations are equivalent (setting aside, for purposes of this discussion, the question whether there may be a mathematical inequivalence of some kind)—even metaphysically equivalent—for they represent the same physical reality, by assumption. However, even given that assumption, there is still some room for denying that they are *completely* metaphysically equivalent, since they are not equally *direct* presentations of reality. For the potentials are—by stipulation!—only indirect descriptions of what's physically real, namely the fields. The two formulations ultimately say all the same things about the world, but one of them plays it straight, the other beats around the bush. So even here, there is a *kind* of inequivalence between them: the two formulations have different degrees of metaphysical directness or transparency.

(The distinction between direct and indirect representations, and the idea that there can be differing degrees of metaphysical directness, only makes sense assuming that a given formulation does not say, of its own predicates, that they are the fundamental ones, representing all and only the basic quantities; for then any other formulation, which says the same thing of its predicates, would be flat-out contradictory, not merely less direct. I am assuming that a formulation does not come with such a claim, that this requires additional interpretive work or physical stipulation. Incidentally, the various subtleties being discussed here, such as a theory's initial physical posits and the possibility of differing degrees of directness, is why I mentioned, in Chapter 1, that figuring out which formulation is most direct is subtle and complicated.)

We could say the same thing of the different formulations of traditional Newtonian gravitation. We can stipulate from the outset that the theory is fundamentally about gravitational forces, in which case the gravitational potential is just an indirect way of describing the forces, and the Poisson equation just an alternative formulation of the inverse square law. Relative to this stipulation, the two formulations present the same picture of physical reality, and in that way they are metaphysically equivalent. But in another way, they are still not *completely* metaphysically equivalent, since only the force-based formulation—by stipulation—*directly* represents that reality.

A similar thing can be said of the Heisenberg and Schrödinger "pictures" or "representations" of non-relativistic quantum mechanics. Empirically and mathematically equivalent; informationally equivalent. Even so, on one natural understanding, they present different pictures of the physical world. According to the Heisenberg picture, there is one state of the world (or any sub-system) that is unchanging with time; instead, the observables or operators evolve. According to the Schrödinger picture, the physical state of the world (or any sub-system) itself evolves in time. Now here, too, we can stipulate that the Heisenberg representation is just a different, roundabout way of saying all the same things the Schrödinger one does, that one formulation is simply more direct than the other. (This may even be something like the standard view in physics.) Yet even though this ensures that the two are in a sense metaphysically equivalent, it also means that in another sense, they are not *completely* metaphysically equivalent, since only one of them directly represents what is going on physically, by stipulation. (In conversation I have heard people say that the Heisenberg picture is not even an intelligible picture of the world, a view on which it is clearly not metaphysically equivalent to the Schrödinger one.)

Thus, whereas Sklar says, "Hardly anyone would deny that the Schrödinger and Heisenberg 'representations' are, indeed, representations of one and the same theory, despite the fact that in the former the state function varies with time and the operators do not and in the latter the reverse is the case" (1982, 90), there is room for regarding them as distinct theories, for one of two reasons. (Keep in mind that the non-equivalence concerns the two "pictures," or *theories*, not the formalisms alone.) (1) On a natural conception, they are not metaphysically equivalent, for they present different physical realities. (2) Alternatively, stipulate that they do represent the same physical reality; even then, there is a kind of metaphysical inequivalence between them, in that only one of them directly represents that reality.

Something similar can be said of quantum mechanics formulated in the position versus the momentum basis, often seen as a paradigmatic case of theoretical equivalence in physics. These are indeed equivalent in all sorts of ways, and yet depending on your preferred solution to the measurement problem, you might think that only one of them directly represents what is going on physically. According to GRW, for example, the position basis representation is the one that directly describes the true collapse of the wavefunction; according to Bohm's theory, it is the one in which particles' genuine positions are represented. On these theories, the momentum basis formulation is an indirect representation of what is happening physically—even though all the relevant quantum mechanical information is recoverable from the momentum-basis representation by means of a straightforward mathematical transformation. Momentum and position may be mathematically on a par, in other words, but just as in the case of the potentials and the fields in classical electromagnetism, this does not mean that they must be metaphysically on a par as well. (A view such as Wallace and Timpson's, again, will simply disagree.)

Now, whether a different "degree of metaphysical directness" interferes with theories' wholesale equivalence, or whether this is more of a presentational difference, is something I am not entirely sure about. Is there a substantive choice to be made between two formulations that present the same picture of the world, but differ in how directly they do this? On the one hand, it seems like there can't be: they describe the same physical reality, by assumption; they respect the intuitive idea of "saying all the same things, but perhaps in different ways." On the other hand, we generally prefer direct formulations, and with good scientific reason. Directness brings with it a level of metaphysical perspicuousness that can result in theoretical progress, often in ways that are hard to foresee—Heaviside's reformulation of Maxwell's equations being a case in point. (In earlier chapters I suggested that we ought to prefer direct formulations, ceteris paribus, but I did not say that a difference in degree of directness amounts to a distinct theory.) I am not sure what the final answer is, nor whether there must be one. Perhaps we can do no better than to say that in one sense, two such formulations are wholly (metaphysically) equivalent, while in another sense, they are not wholly (metaphysically) equivalent.

Although all of this has been getting increasingly controversial, I want to reiterate the fact that the kinds of considerations I have been drawing on are ones that are of central concern to ordinary physics—considerations having to do with what there is in the physical world, what it is like, how it behaves, and how it explains the observable phenomena.

Let me end this section with a case that may be most controversial of all. Classical mechanics can be formulated in a number of ways, two of which, the Lagrangian and Newtonian formulations, were the focus of Chapter 4. As mentioned earlier, these are generally regarded as empirically and formally equivalent, hence fully equivalent: mere notational variants.

I argued that there is a certain mathematical or structural inequivalence between the two, corresponding to a difference in the structure of physical space. Yet even setting any such differences aside, on a natural understanding, they are not metaphysically equivalent, for reasons we were beginning to see in Chapter 4. They share the assumption of a fundamental ontology of point-particles moving around and interacting in three-dimensional physical space. Beyond that, they contain different pictures of the physical world, built up out of different quantities, with correspondingly different explanations of the phenomena.[34]

[34] Occasionally physicists have suggested that these are not equivalent, for reasons similar to those being discussed here. Hertz describes different versions of mechanics as presenting different "images" of the world, based on different fundamental assumptions, which may furthermore not be epistemically on a par. In his words: "By varying the choice of the propositions which we take as fundamental, we can give various representations of the principles of mechanics. Hence we can thus obtain various images of things; and these images we can test and compare with each other in respect of permissibility, correctness, and appropriateness" (1899, 4). See also the introduction and first chapter of Lanczos (1970).

The Newtonian dynamical law centers on forces. On a natural understanding of Newtonian mechanics (one that seems close to what Newton had in mind[35]), particles move around and interact as a result of the forces acting between them. Forces, which are vector quantities, are real physical things in the world, which exist over and above the particles, and cause and explain their behavior. Newtonian mechanics "describes the world in terms of forces and accelerations (as related by the second law)" (Taylor, 2005, 521), as one textbook puts it.

The Lagrangian dynamical law centers on the Lagrangian, a scalar energy function. On a natural understanding of this theory, particles move around and interact as a result of their energies. Energy is a real physical thing in the world, which exists over and above the particles, and causes and explains their behavior.

Although energy and force functions are mathematically inter-derivable in straightforward ways that physics books will show, remember that being mathematically on a par does not imply being metaphysically on a par; similarly, being mathematically definable using a theory's formalism does not necessarily mean corresponding to something physically real (think of the potentials in classical electromagnetism). And on a natural understanding, these theories present different pictures of the world.

According to Newtonian mechanics, on a natural understanding, force is the fundamental dynamical quantity, with energy a secondary or derivative quantity. Energy can be seen as physically real, but this is not what's ultimately responsible for particles' behavior.[36] When we explain why particles move around and interact in the ways they do, we cite the forces at work and the accelerations they produce. Of course, we often do mention the energy in explaining a phenomenon or solving a problem; sometimes this is even the simplest way to do so. Yet although energy is useful for certain purposes, from the perspective of this theory, there must in principle be an explanation available in terms of inter-particle forces, which cites laws given in terms of these forces—just as in classical electromagnetism, there is in principle an explanation of any phenomenon in terms of fields, which cites laws stated in terms of the fields, even though the potentials can be very useful devices.

According to Lagrangian mechanics, on a natural understanding, energy is the fundamental dynamical quantity that explains how and why particles move around and interact in the ways they do. This is the quantity that features in the laws, which we cite in predicting and explaining particles' behavior. From the perspective of this theory, force is "a secondary quantity" derivable from the energy, rather than being "something primitive and irreducible" as it is in Newtonian

[35] This is not to deny that there are questions about exactly how Newton conceived of forces; see Jammer (1999, Ch. 7).

[36] We might say either that energy is physically real but not fundamental, or that it is merely a mathematically definable quantity not corresponding to anything physical. Compare Maudlin's (2018) "derivative ontology" versus "mathematical fictions."

mechanics (Lanczos, 1970, 27). One textbook goes so far as to say that "nowhere in the Lagrangian formulation does the concept of *force* enter" (Marion and Thornton, 1995, 262). (One sometimes hears the Lagrangian function referred to as nomological, as summarizing the laws. This assumes that a classical world is fundamentally Newtonian and that the Lagrangian theory is merely an alternative mathematical formulation of the Newtonian one. I am discussing a conception of the theories on which they describe fundamentally different worlds—on which energy is part of the physical ontology of the Lagrangian theory, the counterpart to the physically real forces of the Newtonian theory.)

Lagrangian and Newtonian mechanics may be informationally equivalent. We may be able to recover all the relevant physical information or facts from each one (setting aside any reasons to think otherwise from Chapter 4); each is reasonably considered a "theory of classical mechanics." Nonetheless, they are not metaphysically equivalent. The worlds they describe are built up out of different quantities, which enter into different laws, resulting in different explanations of the phenomena. The two formulations are not completely equivalent: they differ in ways that matter to the physics.[37] (As we saw in Chapter 4, Wilson sees these as essentially equivalent theories; yet she herself notes that, "Newtonian forces represent a distinctive level of explanatory unity," for they "unify phenomena in a distinctive way" (2007, 196–7). I agree, and would add that energy- and force-based approaches each explain the phenomena in their own distinctive ways.)[38]

Coffey sees this as another case of asymmetrical equivalence. The two formulations are fully equivalent, yet the Newtonian one is primary, the Lagrangian one a reformulation of it. I think this is instead another case of equivalence in certain respects and not others. On a natural understanding, the theories describe different physical realities, even though they are equivalent in the ways that underlie the standard view—they are informationally but not metaphysically equivalent. Again, it is open to *stipulate* that Lagrangian and Newtonian mechanics describe the same physical reality, so that they are metaphysically equivalent after all; stipulate that the Lagrangian function is simply an alternative, roundabout way of coding up things about Newtonian forces. But this requires a physical stipulation to that effect, which furthermore allows for a kind of inequivalence between the two in terms of their relative directness. Absent such a stipulation, they are not metaphysically equivalent.

[37] We might say that they posit different quantities as being "fundamental in the physicist's sense," as Ruetsche (2011, 31) calls it, meaning that from the given quantities, we can calculate all other relevant magnitudes for any system. (I would broaden this to include the central role the quantities play in explanation and prediction and the laws.) The phrase is intended to allow us to sidestep the question whether these are fundamental in the metaphysician's sense.

[38] The Hamiltonian formulation of classical mechanics treats momentum as a fundamental quantity, not defined in terms of the time derivative of position. This, among other reasons, suggests an inequivalence among all three formulations of classical mechanics, despite their standardly claimed equivalence: see North (2009, forthcoming).

A final twist on the case of classical mechanics. Recall that Newtonian and Lagrangian mechanics can themselves each be formulated mathematically in terms of n points in a three-dimensional space, or a single point in a $3n$-dimensional configuration space (or a $6n$-dimensional statespace). Ordinarily, we take the high-dimensional formulation to be just an alternative mathematical formulation of the theory. It is isomorphic and empirically equivalent to the low-dimensional formulation.

We can now see two ways in which that usual idea is too quick. First, the reason we regard the formulations as equivalent is that we ordinarily assume from the outset that classical mechanics is fundamentally about n particles in a three-dimensional physical space, and the configuration space or statespace is constructed on the basis of this assumption. The high-dimensional formulation is stipulated from the beginning to be about multiple particles moving around in three-dimensional space (which is why the statespace has the dimensionality and structure it does). Bringing this initial assumption to light then reveals *a* way of drawing a non-equivalence between low- and high-dimensional formulations of classical mechanics, which is that only the former *directly* represents what is going on physically, by stipulation. (Why not say that in Lagrangian mechanics, particles are stipulated to be in Euclidean space, contrary to the conclusion of Chapter 4? As I conceived of the theories in that chapter, we do not stipulate the metric of physical space at the outset, but have to learn about it from the dynamics. One could still insist on making such an initial stipulation, though this would result in an epistemically inferior theory, containing more structure than what's required by the dynamics.)

Second, consider a classical world that contains only a single particle moving through a physical space that is isomorphic to what we ordinarily think of as the $3n$-dimensional configuration space (or $6n$-dimensional statespace) for large n, where the particle moves through the high-dimensional space exactly as a point representing the state of a three-dimensional world with n particles would move through its configuration space (or statespace). In light of this world, the high- and low-dimensional formulations of classical mechanics are not fully equivalent, for they are not metaphysically equivalent: only one of them accurately represents physical reality.[39] There is a real metaphysical—or just plain physical—difference between a world with a single particle moving through a high-dimensional physical space, and a world with many particles moving through a three-dimensional physical space—even if the two mathematical formulations are informationally equivalent in the sense that each one can be used to recover or "define up" the physics of either world.

[39] That is, assuming there is no initial stipulation that the three-dimensional formulation is just an alternative mathematical way of saying things about the high-dimensional world!

So there is room for drawing a metaphysical inequivalence even between what we usually think of as ordinary-space versus statespace formulations of classical mechanics: either because they are distinct theories, describing different physical worlds or realities, so that only one of them can be correct; or because only one of them directly represents the physical world.

7.6 Explanation matters

If a physical theory is an account of what there is in the world, what it is like, and how and why it behaves in various ways, then two physical theories cannot be fully equivalent if they differ in what there is, what it is like, or how and why it behaves in various ways—if they are not what I have been calling metaphysically equivalent.

The idea of metaphysical equivalence is not completely precise. There will be room for disagreement over any given case. This doesn't change the fact that these considerations *matter* to our judgments of equivalence in physics. Indeed, I take this to be one of the chief lessons of discussions in the foundations of quantum mechanics over the past decades. In rejecting the orthodox theory as inadequate, for instance, we are saying that more than formal and empirical considerations matter to both the evaluation and identification of a physical theory, so that more than this must also matter to the equivalence between physical theories.

This goes along with a very general idea that the aim of science is not just to describe things, but to *explain* the phenomena, something I have been emphasizing is a familiar thought in ordinary science. As the physicist Sean Carroll puts it, "it's wrong to think of the goal of science as simply to fit the data. The goal of science goes much deeper than that: It's to *understand* the behavior of the natural world," "to *explain* what we see" (2010, 371; 370; original italics). A theory's "picture of the world" is essential to this element of explanation and understanding that we demand from science. This is why I said, back in Chapter 1, that my approach is not metaphysical hubris (as Saatsi would have it), but a basic part of science, ordinarily understood.

I mentioned at the beginning of the chapter that one of my main points should be reasonably uncontroversial, which is that when it comes to the equivalence of *physical* theories, it is hard to see how any formal criterion can suffice, simply because physical theories themselves are not purely formal things. (It is noteworthy that discussions such as Barrett and Halvorson (2016b, 2017) focus on examples from pure mathematics, such as different formulations of Euclidean geometry or group theory.) The formal apparatus of two physical theories can be mathematically equivalent in a sense, even identical, yet the theories can still be saying different things about the physical world. (As van Fraassen says in making a related point, "A representation of gas diffusion is not the same thing as a representation of temperature distribution, even if the math is the same"

(2014, 279).)[40] Nor does adding the requirement of empirical equivalence help enough. Physical theories crucially make claims about the world, which go beyond the observable, are not entailed by the formalism, and yet are essential to their explanations and predictions of the phenomena. Questions about the equivalence of physical theories cannot be assessed independently of those claims. That general point remains regardless of whether you agree with my verdicts on particular cases.

There is another way of seeing the point that a physical theory is more than its formalism, so that the equivalence of physical theories must involve more than their respective formalisms. Consider Barrett and Halvorson's (2017) argument that for any geometry, there will be equivalent formulations available in terms of lines or in terms of points. They show that the formulations will be equivalent in a particular sense called Morita equivalent. The basic idea is that the two theories will have a common "Morita extension," a "larger" theory that quantifies over both lines and points. (The Morita extension is arrived at by defining new "sorts" and their associated quantifiers out of the original ones.) Relative to their common Morita extension, each formulation of the geometry will have the resources to define all the vocabulary of the other. In this sense, the two formulations are inter-translatable: they "express the same geometric facts" (Barrett and Halvorson, 2017, 1044).

Barrett and Halvorson go on to say that the theories' having a common Morita extension allows us to think of the original two theories—the one formulated solely in terms of points, and the one formulated solely in terms of lines—as having the same ontological commitments. In particular, we needn't think of the sort symbols and associated quantifiers of the common Morita extension as new to either of the original theories, bringing with them any new ontological commitments. For, "there is a sense in which they were implicitly there in the theory . . . to begin with," so that this "is just a way of making more explicit the ontological commitments of the original theory" (Barrett and Halvorson, 2017, 1059; 1056). The theory formulated in terms of points was in effect already committed to the existence of lines, since lines are definable using the resources of the point-based theory; and the theory formulated in terms of lines was in effect already committed to the existence of points, since points are definable using the resources of the line-based theory.

Although that may be right for purely mathematical theories (I am not concerned here to evaluate the mathematical case), when it comes to scientific theories, there is a difference between their expressing the same facts, on the one

[40] Maxwell, who developed a molecular vortex model to help explain his theory of electromagnetism, was clear about this (Bokulich, 2015). As one example, here is Maxwell on the similarity in form of certain equations for heat conduction and for forces of attraction: "the conduction of heat is supposed to proceed by an action between contiguous parts of a medium, while the force of attraction is a relation between distant bodies, and yet, if we knew nothing more than is expressed in the mathematical formulae, there would be nothing to distinguish between the one set of phenomena and the other" (1890b, 157).

hand, and our being able to indirectly recover the same facts (given suitable definitions), on the other. In classical electromagnetism, the potentials are definable using the mathematical resources of the fields formulation, but this on its own does not show that classical electromagnetism is ontologically committed to the potentials, something that most people will in fact deny. Given the right definitions, we can recover the same physical facts from the potentials and fields formulations. Nonetheless, we do not conclude from this that they present the same picture of the world: informational equivalence does not imply metaphysical equivalence. More generally, as we see in the case of classical electromagnetism, a physical theory is not implicitly committed to the physical reality of everything that can be defined up using its formalism. Again, we may *stipulate* that two such formulations are ultimately about the same physical ontology, presenting the same picture of the world. But this requires the additional physical stipulation, which does not follow from formal considerations alone.

I have not discussed every potential case of theoretical equivalence, nor do I have a verdict for every case one can think of. (Nor have I discussed every respect in which theories may fail to be equivalent. One reason behind Wilson's (2009, 2013, forthcoming) view of the inequivalence of different formulations of classical mechanics, for example, is their differing problem-solving methods.) The important thing to realize is that the considerations I have been discussing do not have to yield an immediate or conclusive verdict on every single case in order to reasonably conclude that the metaphysical aspects of a physical theory matter to the question of its equivalence with other theories, in ways that are relevant to physics as ordinarily understood.

Let me end with two final examples further suggesting that these considerations will be important to any potential case of theoretical equivalence in physics.

First, take the standard mathematical statement of Newton's second law, which is said to hold only in inertial frames, and compare it with a formulation that is applicable to non-inertial frames. Are these equivalent formulations of Newton's law?

The usual view is equivocal. In one way, they are treated as fully equivalent: the second is said to be a reformulation of Newton's law, a way of stating *this very law* so that it applies to non-inertial reference frames. In another way, they are not treated as fully equivalent. The reformulated equation contains additional terms, which appear to correspond to pseudo forces. They are called "force terms" since they look like forces if we interpret the equation so as to preserve the original law's connection between forces and acceleration. But they are also called "pseudo" or "fictitious" since the physical things they represent, if they existed, would not obey the usual Newtonian laws. (They disobey the action–reaction constraint of the third law, for instance.) For that reason, they are said to not correspond to genuine physical forces, nor indeed to anything in physical reality, but are merely mathematical artifacts of having chosen a non-inertial reference frame.

Are these notational variants of a single law? In one way, they are: we have simply restated Newton's law so that it applies to non-inertial frames. In another way, they are not: on a natural understanding, one of them suggests the existence of a kind of force that is prohibited by the original theory.

The considerations discussed in this chapter allow us to see the reason for the equivocation. It depends on what we stipulate about pseudo forces. If we allow for the physical reality of such things, then the second equation is not just a reformulation of the original, but more like (part of) a distinct theory—one that allows for a different kind of physical force—and we may indeed say that Newton's original law holds only in inertial reference frames. (Calling them "pseudo forces" might seem to suggest that we ought not adopt such a view, but allowing for the physical reality of such things is not completely beyond the pale. In the relevant frame, the given force will seem as real as any third-law-obeying force, it will be used to explain and predict the phenomena in that frame: just think of the force you feel as your car accelerates around a bend in the road, or the Coriolis force due to the rotation of the earth.[41])

If we instead stipulate that there is nothing in physical reality corresponding to those additional mathematical terms, then the restated equation *is* a mere reformulation, and the additional terms are indeed mere artifacts of formulation—though at the same time, the reformulated equation is in that case less direct, and is in that sense perhaps not a *mere* reformulation. Notice that, given this stipulation about the pseudo force terms, Newton's law does apply to non-inertial frames after all, albeit not in its original form. (Recall from Chapter 3 that we should say that Newton's law *in its original form* doesn't apply to non-inertial frames, not that Newton's law, full stop, doesn't apply.) The usual view is equivocal about the reformulated equation because the usual view is unclear about the different (meta)physical commitments one can make.

(Similar things can be said of the covariant formulation of Newton's second law, mentioned in Chapter 4, also often said to be just an alternative formulation of Newton's law. The covariant form of the equation, too, contains additional terms, giving rise to similar questions. In this case, it will seem odder to allow for the

[41] Compare the point from Chapter 2 that frame-dependent quantities needn't be completely unreal. Although pseudo forces are usually said to be wholly physically unreal by physics books, the occasional book suggests otherwise, for the above sorts of reasons. An example: "Whenever the motion of the reference system generates a force which has to be added to the relative force of inertia I', measured in that system, we call that force an 'apparent [fictitious] force'. The name is well chosen, inasmuch as that force does not exist in the absolute system and is created solely by the fact that our reference system moves relative to the absolute system. The name is misleading, however, if it is interpreted as a force which is not as 'real' as any given physical force. In the moving reference system the apparent force is a perfectly real force which is not distinguishable in its nature from any other impressed force.... If the physicist who is unaware of his own motion interprets the apparent force... as an external force, he comes into no conflict with the facts" (Lanczos, 1970, 98). Another says: "One can take the view that the inertial force is a 'fictitious' force, introduced merely to preserve the form of Newton's second law. Nevertheless, for an observer in an accelerating frame, it is entirely real" (Taylor, 2005, 329).

physical reality of the things represented by these terms, since they would lack some of the defining features of "ordinary" pseudo forces, such as going to zero in inertial frames. The point remains that there are different options for how to understand the reformulated law, relative to different physical assumptions, so that it is not completely clear-cut whether to consider it merely a notational variant of the very same law.)

As a final example, consider general relativity formulated in terms of differentiable manifolds versus Einstein algebras. Sarita Rosenstock, Thomas Barrett, and James Weatherall (2015) argue that these are empirically and formally equivalent, and conclude that they are therefore wholly equivalent. However, there is a natural understanding on which they are not metaphysically equivalent. The traditional substantivalist, for one, who believes in a manifold of spacetime points, will likely think that only one of them accurately represents the nature of spacetime. The two formulations may be informationally equivalent; all the general relativistic facts may be recoverable from each one; nonetheless, they are not wholly equivalent, because they are not metaphysically equivalent. Here, once again, there is the option of stipulating that both depict the true nature of spacetime, even if only one of them does so directly. But this requires the additional stipulation, which at the same time drives a wedge between the two formulations in terms of their relative directness. There is even the option of stipulating that they are equally good ways of representing the nature of spacetime, neither one more direct, as on the Bainian view mentioned in Chapter 5. (A kind of quotienting view.) This, too, requires an additional stipulation to that effect. Wholesale equivalence cannot be deduced from a combination of formal and empirical equivalence alone.

You may wonder whether I allow for there to be any cases of fully equivalent theoretical formulations. I do, as for example Lagrangian mechanics stated in terms of different types of coordinates, or Newtonian mechanics stated in terms of the coordinates of different inertial frames. Yet it is also true that I think there are many fewer cases of wholesale equivalence in physics than usually thought. I do not see this as a problem. I can still capture what is of significance behind familiar claims of equivalence in physics, by focusing on the various respects in which theories are, or are not, equivalent to one another. I can allow that the geometrized formulation of Newtonian gravity yields insight into the traditional theory and its relationship to general relativity, for example, given the respects in which the geometrized theory is similar to both general relativity and the traditional formulation, even if none of these are fully equivalent physical theories. Given this, it is hard to see what my view is missing, while at the same time it allows us to be alert to potentially far-reaching theoretical differences being glossed over by formal accounts.

Feynman once wrote that, "every theoretical physicist who is any good knows six or seven different theoretical representations for exactly the same physics. He knows that they are all equivalent, and that nobody is ever going to be able to decide which one is right at that level" (1965, 168). The different representations may be useful for different purposes, but there is no real choice to be made between them. This is a familiar thought. The problem with this thought is that there are different senses in which formulations can be said to "represent exactly the same physics." Different theories or formulations can represent the same physics in the sense that we can recover the same physical facts from each one. Yet at the same time, they can fail to represent the same physics in that they fail to depict the same kind of physical reality. In other words, informational equivalence does not suffice for wholesale equivalence in physics. Metaphysical—or just plain physical—considerations matter too.

References

Aharonov, Y. and D. Bohm (1959). "Significance of Electromagnetic Potentials in the Quantum Theory." *Physical Review* 115, 485–91.

Albert, David Z (1996). "Elementary Quantum Metaphysics." In James T. Cushing, Arthur Fine, and Sheldon Goldstein, eds., *Bohmian Mechanics and Quantum Theory: An Appraisal*, pp. 277–84. Dordrecht: Kluwer Academic Publishers.

Albert, David Z (2000). *Time and Chance*. Cambridge, Mass.: Harvard University Press.

Albert, David Z (2015). "Quantum Mechanics and Everyday Life." In *After Physics*, chap. 8. Cambridge, Mass.: Harvard University.

Albert, David Z (2018). "Philosophy of Physics." In *Encyclopaedia Britannica* https://www.britannica.com/topic/philosophy-of-physics.

Albert, David Z (2019a). "How to Teach Quantum Mechanics." Unpublished manuscript. Available at http://philsci-archive.pitt.edu/15584/.

Albert, David Z (2019b). "Preliminary Considerations on the Emergence of Space and Time." In Alberto Cordero, ed., *Philosophers Look at Quantum Mechanics*, pp. 87–95. Switzerland: Springer.

Albert, David Z (2020). "On the Emergence of Space and Time." Unpublished manuscript.

Allori, Valia (2013). "Primitive Ontology and the Structure of Fundamental Physical Theories." In Alyssa Ney and David Z Albert, eds., *The Wave Function: Essays on the Metaphysics of Quantum Mechanics*, pp. 58–75. Oxford: Oxford University Press.

Allori, Valia (2015a). "Maxwell's Paradox: the Metaphysics of Classical Electrodynamics and its Time-Reversal Invariance." *Analytica* 1, 1–19.

Allori, Valia (2015b). "Primitive Ontology in a Nutshell." *International Journal of Quantum Foundations* 1, 107–22.

Allori, Valia (2019). "Quantum Mechanics, Time and Ontology." *Studies in History and Philosophy of Modern Physics* 66, 145–54.

Allori, Valia, Sheldon Goldstein, Roderich Tumulka, and Nino Zanghì (2008). "On the Common Structure of Bohmian Mechanics and the Ghirardi-Rimini-Weber Theory." *British Journal for the Philosophy of Science* 59, 353–89.

Anderson, James L. (1967). *Principles of Relativity Physics*. New York: Academic Press.

Arnold, V. I. (1989). *Mathematical Methods of Classical Mechanics*. New York: Springer, 2nd edn. Translated by K. Vogtmann and A. Weinstein.

Arnold, V. I. and A. B. Givental (2001). "Symplectic Geometry." In V. I. Arnold and S. P. Novikov, eds., *Dynamical Systems IV: Symplectic Geometry and its Applications*, 5–138. Berlin: Springer-Verlag, 2nd edn. Translated by G. Wasserman.

Arntzenius, Frank (1995). "Indeterminism and the Direction of Time." *Topoi* 14, 67–81.

Arntzenius, Frank (1997). "Mirrors and the Direction of Time." *Philosophy of Science (Proceedings)* 64, S213–22.

Arntzenius, Frank (2012). *Space, Time, and Stuff*. Oxford: Oxford University Press.

Arntzenius, Frank and Cian Dorr (2012). "Calculus as Geometry." In *Space, Time, and Stuff*, chap. 8. Oxford: Oxford University Press.

Baez, John C. and Derek K. Wise (2005). "Lectures on Classical Mechanics." https://math.ucr.edu/home/baez/classical/.

Bain, Jonathan (2006). "Spacetime Structuralism." In Dennis Dieks, ed., *The Ontology of Spacetime*, pp. 37–65. Amsterdam: Elsevier.

Baker, David John (2005). "Spacetime Substantivalism and Einstein's Cosmological Constant." *Philosophy of Science (Proceedings)* 72, 1299–311.

Baker, David John (2010). "Symmetry and the Metaphysics of Physics." *Philosophy Compass* 5: 1157–66.

Baker, David John (2020). "Knox's Inertial Spacetime Functionalism (and a Better Alternative)." *Synthese*. https://doi.org/10.1007/s11229-020-02598-z.

Barbour, Julian (1982). "Relational Concepts of Space and Time." *British Journal for the Philosophy of Science* 33, 251–74.

Barbour, Julian (1999). *The End of Time: The Next Revolution in Physics*. New York: Oxford University Press.

Barbour, Julian and Bruno Bertotti (1982). "Mach's Principle and the Structure of Dynamical Theories." *Proceedings of the Royal Society of London. Series A, Mathematical and Physical Sciences* 382, 295–306.

Barrett, Jeffrey Alan (2008). "Approximate Truth and Descriptive Nesting." *Erkenntnis* 68, 213–24.

Barrett, Thomas William (2015a). "On the Structure of Classical Mechanics." *British Journal for the Philosophy of Science* 66, 801–28.

Barrett, Thomas William (2015b). "Spacetime Structure." *Studies in History and Philosophy of Modern Physics* 51, 37–43.

Barrett, Thomas William (2018). "What Do Symmetries Tell Us About Structure?" *Philosophy of Science* 85, 617–39.

Barrett, Thomas William (2019). "Equivalent and Inequivalent Formulations of Classical Mechanics." *British Journal for the Philosophy of Science* 70, 1167–99.

Barrett, Thomas William (2020a). "How to Count Structure." *Noûs*. https://doi.org/10.1111/nous.12358.

Barrett, Thomas William (2020b). "Structure and Equivalence." *Philosophy of Science* 87: 1184–96.

Barrett, Thomas William and Hans Halvorson (2016a). "Morita Equivalence." *Review of Symbolic Logic* 9, 556–82.

Barrett, Thomas William and Hans Halvorson (2016b). "Glymour and Quine on Theoretical Equivalence." *Journal of Philosophical Logic* 45, 467–83.

Barrett, Thomas William and Hans Halvorson (2017). "From Geometry to Conceptual Relativity." *Erkenntnis* 82, 1043–63.

Beem, John K., Paul E. Ehrlich, and Kevin L. Easley (1996). *Global Lorentzian Geometry*. Boca Raton: Taylor & Francis, 2nd edn.

Bell, J. S. (1987a). "How to Teach Special Relativity." In *Speakable and Unspeakable in Quantum Mechanics*, pp. 67–80. Cambridge: Cambridge University Press.

Bell, J. S. (1987b). *Speakable and Unspeakable in Quantum Mechanics*. Cambridge: Cambridge University Press.

Bell, J. S. (1990). "Against 'Measurement'." In Arthur I. Miller, ed., *Sixty-Two Years of Uncertainty: Historical, Philosophical, and Physical Inquiries into the Foundations of Quantum Mechanics*, pp. 17–31. New York: Plenum.

Bell, J. S. (2004). *Speakable and Unspeakable in Quantum Mechanics*. Cambridge: Cambridge University Press, 2nd edn.

Belot, Gordon (1998). "Understanding Electromagnetism." *British Journal for the Philosophy of Science* 49, 531–55.

Belot, Gordon (1999). "Rehabilitating Relationalism." *International Studies in the Philosophy of Science* 13, 35–52.

Belot, Gordon (2000). "Geometry and Motion." *British Journal for the Philosophy of Science* 51, 561–95.

Belot, Gordon (2003). "Notes on Symmetries." In Katherine Brading and Elena Castellani, eds., *Symmetries in Physics: Philosophical Reflections*, pp. 393–412. Cambridge: Cambridge University Press.

Belot, Gordon (2011). *Geometric Possibility*. Oxford: Oxford University Press.

Belot, Gordon (2013). "Symmetry and Equivalence." In Robert Batterman, ed., *The Oxford Handbook of Philosophy of Physics*, pp. 318–39. Oxford: Oxford University Press.

Belot, Gordon and John Earman (2001). "Pre-Socratic Quantum Gravity." In Craig Callender and Nick Huggett, eds., *Physics Meets Philosophy at the Planck Scale*, pp. 213–55. Cambridge: Cambridge University Press.

Bishop, Richard L. and Samuel I. Goldberg (1980). *Tensor Analysis on Manifolds*. New York: Dover.

Bokulich, Alisa (2015). "Maxwell, Helmholtz, and the Unreasonable Effectiveness of the Method of Physical Analogy." *Studies in History and Philosophy of Science* 50, 28–37.

Bokulich, Alisa (2020). "Losing Sight of the Forest for the Ψ: Beyond the Wavefunction Hegemony." In *Scientific Realism and the Quantum*, Steven French and Juha Saatsi, eds., pp. 185–211. Oxford: Oxford University Press.

Brading, Katherine and Elena Castellani, eds. (2003). *Symmetries in Physics: Philosophical Reflections*. Cambridge: Cambridge University Press.

Brading, Katherine and Elena Castellani (2007). "Symmetries and Invariances in Classical Physics." In Jeremy Butterfield and John Earman, eds., *Handbook of the Philosophy of Science: Philosophy of Physics, Part B*, pp. 1331–67. Amsterdam: Elsevier.

Bradley, Clara (2019). "The Non-Equivalence of Einstein and Lorentz." *British Journal for the Philosophy of Science*. doi:10.1093/bjps.axz014.

Bricker, Phillip (1993). "The Fabric of Space." In Peter A. French, Theodore E. Uehling, Jr., and Howard K. Wettstein, eds., *Midwest Studies in Philosophy*, vol. 18, pp. 271–94. Notre Dame: Notre Dame University Press.

Brighouse, Carolyn (1994). "Spacetime and Holes." *Philosophy of Science (Proceedings)* 1, 117–25.

Brighouse, Carolyn (1999). "Incongruent Counterparts and Modal Relationism." *International Studies in the Philosophy of Science* 13, 53–68.

Brighouse, Carolyn (2014). "Geometric Possibility—An Argument from Dimension." *European Journal for Philosophy of Science* 4, 31–54.

Brown, Harvey R. (2005). *Physical Relativity: Space-Time Structure from a Dynamical Perspective*. Oxford: Oxford University Press.

Brown, Harvey R. and Oliver Pooley (2006). "Minkowski Space-Time: A Glorious Non-Entity." In Dennis Dieks, ed., *The Ontology of Spacetime*, pp. 67–89. Amersterdam: Elsevier.

Brown, Harvey R. and James Read (forthcoming). "The Dynamical Approach to Spacetime Theories." In Eleanor Knox and Alastair Wilson, eds., *The Routledge Companion to Philosophy of Physics*. Routledge. Available at http://philsci-archive.pitt.edu/14592/.

Burns, Keith and Marian Gidea (2005). *Differential Geometry and Topology With a View to Dynamical Systems*. Boca Raton: Chapman & Hall.

Butterfield, Jeremy (2004). "Between Laws and Models: Some Philosophical Morals of Lagrangian Mechanics." Unpublished manuscript. Available at http://philsci-archive.pitt.edu/1937/.

Butterfield, Jeremy (forthcoming). "On Dualities and Equivalences Between Physical Theories." In Christian Wüthrich, Baptiste Le Bihan, and Nick Huggett, eds., *Philosophy Beyond Spacetime*. Oxford University Press.

Callender, Craig (1995). "The Metaphysics of Time Reversal: Hutchison on Classical Mechanics." *British Journal for the Philosophy of Science* 46, 331–40.

Callender, Craig (1999). "Reducing Thermodynamics to Statistical Mechanics: The Case of Entropy." *Journal of Philosophy* 96, 348–73.

Callender, Craig (2000). "Is Time 'Handed' in a Quantum World?" *Proceedings of the Aristotelian Society* 100, 247–69.

Carroll, Sean (2004). *Spacetime and Geometry: An Introduction to General Relativity*. San Francisco: Addison-Wesley.

Carroll, Sean (2010). *From Eternity to Here: The Quest for the Ultimate Theory of Time*. New York: Dutton.

Cartwright, Nancy (1999). *The Dappled World: A Study of the Boundaries of Science*. Cambridge: Cambridge University Press.

Chen, Eddy Keming (forthcoming). "Fundamental Nomic Vagueness." *Philosophical Review*. Available at http://philsci-archive.pitt.edu/18734/1/Nomic%20Vagueness_5.pdf.

Coffey, Kevin (2014). "Theoretical Equivalence as Interpretative Equivalence." *British Journal for the Philosophy of Science* 65, 821–44.

Crampin, M. and F. A. E. Pirani (1986). *Applicable Differential Geometry*. Cambridge: Cambridge University Press.

Creary, Lewis G. (1981). "Causal Explanation and the Reality of Natural Component Forces." *Pacific Philosophical Quarterly* 62, 148–57.

Curiel, Erik (2014). "Classical Mechanics Is Lagrangian; It Is Not Hamiltonian." *British Journal for the Philosophy of Science* 65, 269–321.

Curiel, Erik (2017). "A Primer on Energy Conditions." In Dennis Lehmkuhl, Gregor Schiemann, and Erhard Scholz, eds., *Towards a Theory of Spacetime Theories*, pp. 43–104. New York: Birkhäuser.

Curiel, Erik (2018). "On the Existence of Spacetime Structure." *British Journal for the Philosophy of Science* 69, 447–83.

Dainton, Barry (2010). *Time and Space*. Montreal & Kingston: McGill-Queen's University Press, 2nd edn.

Dasgupta, Shamik (2009). "Individuals: An Essay in Revisionary Metaphysics." *Philosophical Studies* 145, 35–67.

Dasgupta, Shamik (2011). "The Bare Necessities." *Philosophical Perspectives* 25, 115–60.

Dasgupta, Shamik (2016). "Symmetry as an Epistemic Notion (Twice Over)." *British Journal for the Philosophy of Science* 67, 837–78.

de Haro, Sebastian (2017). "Dualities and Emergent Gravity: Gauge/Gravity Duality." *Studies in History and Philosophy of Modern Physics* 59, 109–25.

de Haro, Sebastian (2020). "Spacetime and Physical Equivalence." In Nick Huggett, Keizo Matsubara, and Christian Wüthrich, eds., *Beyond Spacetime: The Foundations of Quantum Gravity*, pp. 257–83. Cambridge: Cambridge University Press.

de Haro, Sebastian and Jeremy Butterfield (2018). "A Schema for Duality, Illustrated by Bosonization." In Joseph Kouneiher, ed., *Foundations of Mathematics and Physics One Century After Hilbert: New Perspectives*, pp. 305–76. Switzerland: Springer.

de Haro, Sebastian, Nicholas Teh, and Jeremy N. Butterfield (2017). "Comparing Dualities and Gauge Symmetries." *Studies in History and Philosophy of Modern Physics* 59, 68–80.

Debs, Talal A. and Michael L. G. Redhead (2007). *Objectivity, Invariance, and Convention: Symmetry in Physical Science*. Cambridge, Mass.: Harvard University Press.

Dewar, Neil (2015). "Symmetries and the Philosophy of Language." *Studies in History and Philosophy of Modern Physics* 52, 317–27.

Dewar, Neil (2019). "Sophistication about Symmetries." *British Journal for the Philosophy of Science* 70, 485–521.

DiSalle, Robert (1994). "On Dynamics, Indiscernibility, and Spacetime Ontology." *British Journal for the Philosophy of Science* 45, 265–87.

DiSalle, Robert (2002). "Newton's Philosophical Analysis of Space and Time." In *The Cambridge Companion to Newton*, I. Bernard Cohen and George E. Smith, eds., pp. 33–56. Cambridge: Cambridge University Press.

Dorato, Mauro (2000). "Substantivalism, Relationism, and Structural Spacetime Realism." *Foundations of Physics* 30, 1605–28.

Dorato, Mauro (2008). "Is Structural Spacetime Realism Relationism in Disguise? The Supererogatory Nature of the Substantivalism/Relationism Debate." In Dennis Dieks, ed., *The Ontology of Spacetime II*, pp. 17–37. Amsterdam: Elsevier Science.

Dorr, Cian (2010). "Of Numbers and Electrons." *Proceedings of the Aristotelian Society* 110, 133–81.

Dorr, Cian (2011). "Physical Geometry and Fundamental Metaphysics." *Proceedings of the Aristotelian Society* 111, 135–59.

Dürr, Detlef, Sheldon Goldstein, and Nino Zanghí (1992). "Quantum Equilibrium and the Origin of Absolute Uncertainty." *Journal of Statistical Physics* 67, 843–908.

Dürr, Detlef, Sheldon Goldstein, and Nino Zanghí (2020). "Quantum Motion on Shape Space and the Gauge Dependent Emergence of Dynamics and Probability in Absolute Space and Time." *Journal of Statistical Physics* 180: 92–134.

Earman, John (1969). "The Anisotropy of Time." *Australasian Journal of Philosophy* 47, 273–95.

Earman, John (1986). *A Primer on Determinism*. Dordrecht: D. Reidel.

Earman, John (1989). *World Enough and Space-Time*. Cambridge, Mass.: MIT Press.

Earman, John (1993). "Underdetermination, Realism, and Reason." *Midwest Studies in Philosophy* 18, 19–38.

Earman, John (2002). "What Time Reversal Invariance Is and Why It Matters." *International Studies in the Philosophy of Science* 16, 245–64.

Earman, John and John Norton (1987). "What Price Spacetime Substantivalism? The Hole Story." *British Journal for the Philosophy of Science* 38, 515–25.

Einstein, Albert (1950). "The Theory of Relativity." In *The Theory of Relativity and Other Essays*, pp. 5–12. New York: MJF Books.

Einstein, Albert (2002). *The Collected Papers of Albert Einstein*, vol. 7. Princeton: Princeton University Press. Translated by Alfred Engel.

Einstein, A., B. Podolsky, and N. Rosen (1935). "Can Quantum-Mechanical Description of Physical Reality Be Considered Complete?" *Physical Review* 47, 777–80.

Emery, Nina (2019). "Actualism without Presentism? Not by way of the Relativity Objection." *Noûs* 53, 963–86.

Esfeld, Michael (2019). "Against the Disappearance of Spacetime in Quantum Gravity." *Synthese*. https://doi.org/10.1007/s11229-019-02168-y.

Esfeld, Michael and Vincent Lam (2008). "Moderate Structural Realism about Space-Time." *Synthese* 160, 27–46.

Fetter, Alexander L. and John Dirk Walecka (2003). *Theoretical Mechanics of Particles and Continua*. New York: Dover.

Feynman, Richard (1965). *The Character of Physical Law*. Cambridge, Mass.: MIT Press.

Feynman, Richard P., Robert B. Leighton, and Matthew Sands (2010). *The Feynman Lectures on Physics: New Millennium Edition*, vol. 1. New York: Basic Books.

Field, Hartry (1980). *Science Without Numbers*. Oxford: Blackwell.
Field, Hartry (1985). "Can We Dispense With Spacetime?" *Proceedings of the 1984 Biennial Meeting of the Philosophy of Science Association* 2, 33–90.
Field, Hartry (1989). *Realism, Mathematics and Modality*. Oxford: Blackwell.
Fine, Kit (2001). "The Question of Realism." *Philosopher's Imprint* 1, 1–30.
Fine, Kit (2012). "Guide to Ground." In Fabrice Correia and Benjamin Schnieder, eds., *Metaphysical Grounding: Understanding the Structure of Reality*, pp. 37–80. Cambridge: Cambridge University Press.
Fock, V. (1964). *The Theory of Space, Time and Gravitation*. Oxford: Pergamon Press, 2nd revised edn.
Freedman, Michael H. and Laurence R. Taylor (1986). "A Universal Smoothing of Four-Space." *Journal of Differential Geometry* 24, 69–78.
French, Steven (2014). *The Structure of the World: Metaphysics and Representation*. Oxford: Oxford University Press.
Friedman, Michael (1983). *Foundations of Space-Time Theories*. Princeton: Princeton University Press.
Gallavotti, Giovanni (1983). *The Elements of Mechanics*. New York: Springer-Verlag.
Geroch, Robert (1978). *General Relativity from A to B*. Chicago: University of Chicago Pres.
Geroch, Robert (1985). *Mathematical Physics (Chicago Lectures in Physics)*. Chicago: University of Chicago Press.
Ghirardi, G. C., A. Rimini, and T. Weber (1986). "Unified Dynamics for Microscopic and Macroscopic Systems." *Physical Review* D 34, 470–91.
Ghirardi, G. C., R. Grassi, and F. Benatti (1995). "Describing the Macroscopic World: Closing the Circle within the Dynamical Reduction Program." *Foundations of Physics* 25, 5–38.
Glymour, Clark (1970). "Theoretical Realism and Theoretical Equivalence." *Philosophy of Science (Proceedings)* 1970, 275–88.
Glymour, Clark (1977). "The Epistemology of Geometry." *Noûs* 11, 227–51.
Glymour, Clark (1980). *Theory and Evidence*. Princeton: Princeton University Press.
Goldstein, Herbert, Charles Poole, and John Safko (2004). *Classical Mechanics*. Reading, Mass.: Pearson Education, 3rd edn.
Goodman, Nelson (1955). *Fact, Fiction, and Forecast*. Cambridge, Mass.: Harvard University Press, 4th edn.
Greaves, Hilary (2011). "In Search of (Spacetime) Structuralism." *Philosophical Perspectives* 25, 189–204.
Greaves, Hilary and David Wallace (2014). "Empirical Consequences of Symmetries." *British Journal for the Philosophy of Science* 65, 59–89.
Halvorson, Hans (2012). "What Scientific Theories Could Not Be." *Philosophy of Science* 79, 183–206.
Halvorson, Hans (2016). "Scientific Theories." In Paul Humphreys, ed., *The Oxford Handbook of Philosophy of Science*, pp. 585–608. New York: Oxford University Press.
Halvorson, Hans (2019). *The Logic in Philosophy of Science*. Cambridge: Cambridge University Press.
Halvorson, Hans and Dimitris Tsementzis (2017). "Categories of Scientific Theories." In Elaine Landry, ed., *Categories for the Working Philosopher*, pp. 402–29. Oxford: Oxford University Press.
Hamill, Patrick (2014). *A Student's Guide to Lagrangians and Hamiltonians*. Cambridge: Cambridge University Press.

Hand, Louis N. and Janet D. Finch (1998). *Analytical Mechanics*. Cambridge: Cambridge University Press.

Hartle, James B. (2003). *Gravity: An Introduction to Einstein's General Relativity*. San Francisco: Addison-Wesley.

Heaviside, Oliver (1892). *Electrical Papers*, vol. 2. London and New York: Macmillan and Co.

Heaviside, Oliver (1893). *Electromagnetic Theory*, vol. 1. London: The Electrician Printing and Publishing Company.

Hertz, Heinrich (1899). *The Principles of Mechanics Presented in a New Form*. London: Macmillan and Co.

Hicks, Michael Townsen and Jonathan Schaffer (2017). "Derivative Properties in Fundamental Laws." *British Journal for the Philosophy of Science* 68, 411–50.

Hirsch, Eli (2011). *Quantifier Variance and Realism: Essays in Metaontology*. Oxford: Oxford University Press.

Hoefer, Carl (1996). "The Metaphysics of Space-Time Substantivalism." *Journal of Philosophy* 93, 5–27.

Hoefer, Carl (1998). "Absolute versus Relational Spacetime: For Better or Worse, the Debate Goes On." *British Journal for the Philosophy of Science* 49, 451–67.

Hohensee, Michael A., Brian Estey, Paul Hamilton, Anton Zeilinger, and Holger Müller (2012). "Force-Free Gravitational Redshift: Proposed Gravitational Aharonov–Bohm Experiment." *Physical Review Letters* 108, 230404.

Horwich, Paul (1978). "On the Existence of Time, Space and Space-Time." *Noûs* 12, 397–419.

Horwich, Paul (1987). *Asymmetries in Time: Problems in the Philosophy of Science*. Cambridge, Mass.: MIT Press.

Hudetz, Laurenz (2019). "Definable Categorical Equivalence." *Philosophy of Science* 86, 47–75.

Huggett, Nick (1999). *Space from Zeno to Einstein: Classic Readings with a Contemporary Commentary*. Cambridge, Mass.: MIT Press.

Huggett, Nick (2006). "The Regularity Account of Relational Spacetime." *Mind* 115, 41–73.

Huggett, Nick and Carl Hoefer (2009). "Absolute and Relational Theories of Space and Motion." *Stanford Encyclopedia of Philosophy (Fall 2009 Edition)*.

Huggett, Nick and Christian Wüthrich (forthcoming). *Out of Nowhere: The Emergence of Spacetime in Quantum Theories of Gravity*. Oxford University Press.

Isham, Chris J. (1994). "Prima Facie Questions in Quantum Gravity." In J. Ehlers and H. Friedrich, eds., *Canonical Gravity: From Classical to Quantum*, pp. 1–21. Berlin: Springer-Verlag.

Isham, Chris J. (2003). *Modern Differential Geometry for Physicists*. New Jersey: World Scientific Lecture Notes in Physics, 2nd edn.

Ismael, Jenann (2016). "From Physical Time to Human Time." In Yuval Dolev and Michael Roubach, eds., *Cosmological and Psychological Time*, pp. 107–24. Switzerland: Springer.

Ismael, Jenann (2020). "What Entanglement Might Be Telling Us: Space, Quantum Mechanics, and Bohm's Fish Tank." In David Glick, George Darby, and Anna Marmodoro, eds., *The Foundation of Reality: Fundamentality, Space, and Time*, pp. 139–53. Oxford: Oxford University Press.

Ismael, Jenann and Bas C. van Fraassen (2003). "Symmetry as a Guide to Superfluous Theoretical Structure." In Katherine Brading and Elena Castellani, eds., *Symmetries in Physics: Philosophical Reflections*, pp. 371–92. Cambridge: Cambridge University Press.

Jammer, Max (1999). *Concepts of Force*. Mineola, New York: Dover.

Janssen, Michel (2002a). "COI Stories: Explanation and Evidence in the History of Science." *Perspectives on Science* 10, 457–522.

Janssen, Michel (2002b). "Reconsidering a Scientific Revolution: The Case of Einstein versus Lorentz." *Physics in Perspective* 4, 421–46.

Janssen, Michel (2009). "Drawing the Line between Kinematics and Dynamics in Special Relativity." *Studies in History and Philosophy of Modern Physics* 40, 26–52.

Jeffreys, Harold (1931). *Cartesian Tensors*. Cambridge: Cambridge University Press.

Jones, Roger (1991). "Realism about What?" *Philosophy of Science* 58, 185–202.

José, Jorge V. and Eugene J. Saletan (1998). *Classical Dynamics: A Contemporary Approach*. Cambridge: Cambridge University Press.

Klein, Felix (1892). "A Comparative Review of Recent Researches in Geometry." *Bulletin of the New York Mathematical Society* 2, 215–49. Translated by M. W. Haskell. Available at http://arxiv.org/abs/0807.3161.

Knox, Eleanor (2011). "Newton-Cartan Theory and Teleparallel Gravity: The Force of a Formulation." *Studies in History and Philosophy of Modern Physics* 42, 264–75.

Knox, Eleanor (2013). "Effective Spacetime Geometry." *Studies in History and Philosophy of Modern Physics* 44, 346–56.

Knox, Eleanor (2014). "Newtonian Spacetime Structure in Light of the Equivalence Principle." *British Journal for the Philosophy of Science* 65, 863–80.

Knox, Eleanor (2019). "Physical Relativity from a Functionalist Perspective." *Studies in History and Philosophy of Modern Physics* 67, 118–24.

Ladyman, James (1998). "What is Structural Realism?" *Studies in History and Philosophy of Science* 29, 409–24.

Ladyman, James (2016). "Structural Realism." *Stanford Encyclopedia of Philosophy (Winter 2016 Edition)*.

Ladyman, James and Don Ross (2009). *Every Thing Must Go: Metaphysics Naturalized*. Oxford: Oxford University Press.

Lanczos, Cornelius (1970). *The Variational Principles of Mechanics*. New York: Dover, 4th edn.

Landau, L. D. and E. M. Lifshitz (1976). *Mechanics*. Oxford: Butterworth-Heinemann, 3rd edn.

Lange, Marc (2001). "The Most Famous Equation." *Journal of Philosophy* 98, 219–38.

Lange, Marc (2017). *Because Without Cause: Non-Causal Explanations in Science and Mathematics*. New York: Oxford University Press.

Lazarovici, Dustin (2018). "Against Fields." *European Journal for Philosophy of Science* 8, 145–70.

Lee, John M. (2003). *Introduction to Smooth Manifolds*. New York: Springer-Verlag.

Leeds, Stephen (1995). "Holes and Determinism: Another Look." *Philosophy of Science* 62, 425–37.

Lewis, David (1973). *Counterfactuals*. Oxford: Blackwell.

Lewis, David (1983). "Extrinsic Properties." *Philosophical Studies* 44, 197–200.

Loar, Brian (1981). *Mind and Meaning*. Cambridge: Cambridge University Press.

Loewer, Barry (2001). "From Physics to Physicalism." In Carl Gillet and Barry Loewer, eds., *Physicalism and Its Discontents*, pp. 37–56. Cambridge: Cambridge University Press.

Mac Lane, Saunders (1986). *Mathematics, Form and Function*. New York: Springer-Verlag.

Mach, Ernst (1960). *The Science of Mechanics: A Critical and Historical Account of Its Development*. Open Court. Translated by Thomas J. McCormack.

Malament, David B. (1976). "Review of Lawrence Sklar, *Space, Time, and Spacetime*." *Journal of Philosophy* 73, 306–23.

Malament, David B. (1982). "Review of Hartry Field, *Science Without Numbers: A Defense of Nominalism*." *Journal of Philosophy* 79, 523–34.

Malament, David B. (1986). "Newtonian Gravity, Limits, and the Geometry of Space." In Robert G. Colodny, ed., *From Quarks to Quasars: Philosophical Problems of Modern Physics*, pp. 181–201. Pittsburgh: University of Pittsburgh Press.

Malament, David B. (1995). "Is Newtonian Cosmology Really Inconsistent?" *Philosophy of Science* 62, 489–510.

Malament, David B. (2007). "Classical Relativity Theory." In Jeremy Butterfield and John Earman, eds., *Handbook of the Philosophy of Science: Philosophy of Physics, Part A*, pp. 229–73. Amsterdam: Elsevier.

Malament, David B. (2008). "Norton's Slippery Slope." *Philosophy of Science (Proceedings)* 75, 799–816.

Malament, David B. (2012). *Topics in the Foundations of General Relativity and Newtonian Gravitation Theory*. Chicago: University of Chicago Press.

Manders, Kenneth L. (1982). "On the Space-Time Ontology of Physical Theories." *Philosophy of Science* 49, 575–90.

Marcolli, Matilde (2020). *Lumen Naturae: Visions of the Abstract in Art and Mathematics*. Cambridge, Mass.: MIT Press.

Marion, Jerry B. and Stephen T. Thornton (1995). *Classical Dynamics of Particles and Systems*. Fort Worth: Harcourt Brace, 4th edn.

Marsden, Jerrold E. (1992). *Lectures on Mechanics*. Cambridge: Cambridge University Press.

Maudlin, Tim (1988). "The Essence of Space-Time." *Philosophy of Science (Proceedings)* 2, 82–91.

Maudlin, Tim (1993). "Buckets of Water and Waves of Space: Why Spacetime is Probably a Substance." *Philosophy of Science* 60, 183–203.

Maudlin, Tim (2002). "Thoroughly Muddled McTaggart, or How to Abuse Gauge Freedom to Generate Metaphysical Monstrosities." *Philosopher's Imprint* 2, 1–23.

Maudlin, Tim (2007a). "Completeness, Supervenience and Ontology." *Journal of Physics A: Mathematical and Theoretical* 40, 3151–71.

Maudlin, Tim (2007b). "On the Passing of Time." In *The Metaphysics Within Physics*, pp. 104–42. Oxford: Oxford University Press.

Maudlin, Tim (2010). "Can the World Be Only Wavefunction?" In Simon Saunders, Jonathan Barrett, Adrian Kent, and David Wallace, eds., *Many Worlds? Everett, Quantum Theory, and Reality*, pp. 121–43. Oxford: Oxford University Press.

Maudlin, Tim (2011). *Quantum Non-Locality and Relativity: Metaphysical Intimations of Modern Physics*. Massachussetts: Blackwell, 3rd edn.

Maudlin, Tim (2012). *Philosophy of Physics: Space and Time*. Princeton: Princeton University Press.

Maudlin, Tim (2013). "The Nature of the Quantum State." In *The Wave Function: Essays on the Metaphysics of Quantum Mechanics*, Alyssa Ney and David Z Albert, eds., pp. 126–53. Oxford: Oxford University Press.

Maudlin, Tim (2014a). *New Foundations for Physical Geometry*. Oxford: Oxford University Press.

Maudlin, Tim (2014b). "Critical Study of David Wallace, *The Emergent Multiverse: Quantum Theory According to the Everett Interpretation*." *Noûs* 48, 794–808.

Maudlin, Tim (2015). "How Mathematics Meets the World." http://fqxi.org/data/essay-contest-files/Maudlin_How_Mathematics_Mee.pdf.

Maudlin, Tim (2018). "Ontological Clarity via Canonical Presentation: Electromagnetism and the Aharonov–Bohm Effect." *Entropy* 20, 465.

Maudlin, Tim (2019). *Philosophy of Physics: Quantum Theory*. Princeton: Princeton University Press.

Maxwell, James Clerk (1890a). "On Action at a Distance." In W. D. Niven, ed., *The Scientific Papers of James Clerk Maxwell*, vol. 2, pp. 311–23. New York: Dover.

Maxwell, James Clerk (1890b). "On Faraday's Lines of Force." In W. D. Niven, ed., *The Scientific Papers of James Clerk Maxwell*, vol. 1, pp. 155–229. New York: Dover.

Maxwell, James Clerk (1890c). "On Physical Lines of Force." In W. D. Niven, ed., *The Scientific Papers of James Clerk Maxwell*, vol. 1, pp. 451–513.

Maxwell, James Clerk (1890d). "On the Proof of the Equations of Motion of a Connected System." In W. D. Niven, ed., *The Scientific Papers of James Clerk Maxwell*, vol. 2, pp. 308–9. New York: Dover.

McCauley, Joseph L. (1997). *Classical Mechanics: Transformations, Flows, Integrable and Chaotic Dynamics*. Cambridge: Cambridge University Press.

McKenzie, Kerry (2017). "Ontic Structural Realism." *Philosophy Compass* 12.

McSweeney, Michaela Markham (2016). "An Epistemic Account of Metaphysical Equivalence." *Philosophical Perspectives* 30, 270–93.

Merzbacher, Eugen (1998). *Quantum Mechanics*. New Jersey: John Wiley & Sons, 3rd edn.

Messiah, Albert (2014). *Quantum Mechanics*. Mineola, New York: Dover, 2nd edn.

Muller, F. A. (1997a). "The Equivalence Myth of Quantum Mechanics—Part I." *Studies in History and Philosophy of Modern Physics* 28, 35–61.

Muller, F. A. (1997b). "The Equivalence Myth of Quantum Mechanics—Part II." *Studies in History and Philosophy of Modern Physics* 28, 219–47.

Mundy, Brent (1983). "Relational Theories of Euclidean Space and Minkowski Spacetime." *Philosophy of Science* 50, 205–26.

Mundy, Brent (1986). "Embedding and Uniqueness in Relational Theories of Space." *Synthese* 67, 383–90.

Mundy, Brent (1989). "Distant Action in Classical Electromagnetic Theory." *British Journal for the Philosophy of Science* 40, 39–68.

Mundy, Brent (1992). "Space-Time and Isomorphism." *Philosophy of Science (Proceedings)* 1, 515–27.

Myrvold, Wayne (2019). "How Could Relativity Be Anything Other Than Physical?" *Studies in History and Philosophy of Modern Physics* 67, 137–43.

Nerlich, Graham (1994). *The Shape of Space*. Cambridge: Cambridge University Press, 2nd edn.

Newton, Isaac (1934). *Principia*, vol. 1. Berkeley: University of California Press. Translated by Andrew Motte, revised by Florian Cajori.

Ney, Alyssa (forthcoming). *The World in the Wavefunction*. Oxford: Oxford University Press.

Ney, Alyssa and David Z Albert, eds. (2013). *The Wave Function: Essays on the Metaphysics of Quantum Mechanics*. Oxford: Oxford University Press.

Nguyen, James (2017). "Scientific Representation and Theoretical Equivalence." *Philosophy of Science* 84, 982–95.

Nguyen, James, Nicholas J. Teh, and Laura Wells (2020). "Why Surplus Structure Is Not Superfluous." *British Journal for the Philosophy of Science* 71, 665–95.

Norsen, Travis (2017). *Foundations of Quantum Mechanics: An Exploration of the Physical Meaning of Quantum Theory*. Switzerland: Springer.

North, Jill (2008). "Two Views on Time Reversal." *Philosophy of Science* 75, 201–23.

North, Jill (2009). "The 'Structure' of Physics: A Case Study." *Journal of Philosophy* 106, 57–88.

North, Jill (2013). "The Structure of a Quantum World." In Alyssa Ney and David Z Albert, eds., *The Wave Function: Essays on the Metaphysics of Quantum Mechanics*, pp. 184–202. Oxford: Oxford University Press.

North, Jill (2018). "A New Approach to the Relational–Substantival Debate." *Oxford Studies in Metaphysics* 11, 3–43.

North, Jill (forthcoming). "Formulations of Classical Mechanics." In Eleanor Knox and Alastair Wilson, eds., *The Routledge Companion to Philosophy of Physics*. Routledge.

Norton, John D. (1993a). "General Covariance and the Foundations of General Relativity: Eight Decades of Dispute." *Reports on Progress in Physics* 56, 791–858.

Norton, John D. (1993b). "A Paradox in Newtonian Gravitation Theory." *Philosophy of Science (Proceedings)* 2, 412–20.

Norton, John D. (1995). "The Force of Newtonian Cosmology: Acceleration is Relative." *Philosophy of Science* 62, 511–22.

Norton, John D. (2003). "General Covariance, Gauge Theories, and the Kretschmann Objection." In Katherine Brading and Elena Castellani, eds., *Symmetry in Physics: Philosophical Reflections*, pp. 110–23. Cambridge: Cambridge University Press.

Norton, John D. (2008a). "The Dome: An Unexpectedly Simple Failure of Determinism." *Philosophy of Science (Proceedings)* 75, 786–98.

Norton, John D. (2008b). "Must Evidence Underdetermine Theory?" In Martin Carrier, Don Howard, and Janet Kourany, eds., *The Challenge of the Social and the Pressure of Practice: Science and Values Revisited*, pp. 17–44. Pittsburgh: University of Pittsburgh Press.

Nozick, Robert (1998). "Invariance and Objectivity." *Proceedings and Addresses of the American Philosophical Association* 72, 21–48.

Nozick, Robert (2001). *Invariances: The Structure of the Objective World*. Cambridge, Mass.: Harvard University Press.

Ohanian, Hans C. and Remo Ruffini (2013). *Gravitation and Spacetime*. Cambridge: Cambridge University Press, 3rd edn.

Paul, L. A. (2013). "Categorical Priority and Categorical Collapse." *Aristotelian Society Supplementary Volume* 87, 89–113.

Petersen, Aage (1963). "The Philosophy of Niels Bohr." *Bulletin of the Atomic Scientists* 19, 8–14.

Pooley, Oliver (2013). "Substantivalist and Relationalist Approaches to Spacetime." In Robert Batterman, ed., *The Oxford Handbook of Philosophy of Physics*, pp. 522–86. Oxford: Oxford University Press.

Price, Huw (1996). *Time's Arrow and Archimedes' Point: New Directions for the Physics of Time*. Oxford: Oxford University Press.

Price, Huw (2002a). "Boltzmann's Time Bomb." *British Journal for the Philosophy of Science* 53, 83–119.

Price, Huw (2002b). "Burbury's Last Case: The Mystery of the Entropic Arrow." In Craig Callender, ed., *Time, Reality and Experience*, pp. 19–56. Cambridge: Cambridge University Press.

Quine, W. V. O. (1948). "On What There Is." *Review of Metaphysics* 2, 21–38.

Quine, W. V. O. (1975). "On Empirically Equivalent Systems of the World." *Erkenntnis* 9, 313–28.

Reichenbach, Hans (1958). *The Philosophy of Space and Time*. New York: Dover.

Rickles, Dean and Steven French (2006). "Quantum Gravity Meets Structuralism: Interweaving Relations in the Foundations of Physics." In Dean Rickles, Steven French, and Juha Saatsi, eds., *The Structural Foundations of Quantum Gravity*, pp. 1–39. Oxford: Oxford University Press.

Roberts, John T. (2008). "A Puzzle about Laws, Symmetries and Measurability." *British Journal for the Philosophy of Science* 59, 143–68.

Rosen, Gideon (2010). "Metaphysical Dependence: Grounding and Reduction." In Bob Hale and Aviv Hoffmann, eds., *Modality: Metaphysics, Logic, and Epistemology*, pp. 109–35. Oxford: Oxford University Press.

Rosenstock, Sarita, Thomas William Barrett, and James Owen Weatherall (2015). "On Einstein Algebras and Relativistic Spacetimes." *Studies in History and Philosophy of Modern Physics* 52, 309–16.

Ruetsche, Laura (2011). *Interpreting Quantum Theories: The Art of the Possible.* Oxford: Oxford University Press.

Rynasiewicz, Robert (1996). "Absolute Versus Relational Space-Time: An Outmoded Debate?" *Journal of Philosophy* 93, 279–306.

Rynasiewicz, Robert (2000). "On the Distinction Between Absolute and Relative Motion." *Philosophy of Science* 67, 70–93.

Saatsi, Juha (2019). "Scientific Realism Meets Metaphysics of Quantum Mechanics." In Alberto Cordero, ed., *Philosophers Look at Quantum Mechanics*, pp. 141–62. Switzerland: Springer.

Saunders, Simon (2013). "Rethinking Newton's *Principia*." *Philosophy of Science* 80, 22–48.

Schaffer, Jonathan (2009). "On What Grounds What." In David Chalmers, David Manley, and Ryan Wasserman, eds., *Metametaphysics: New Essays on the Foundations of Ontology*, pp. 347–83. Oxford: Oxford University Press.

Schutz, Bernard F. (1980). *Geometrical Methods of Mathematical Physics.* Cambridge: Cambridge University Press.

Schutz, Bernard F. (2009). *A First Course in General Relativity.* Cambridge: Cambridge University Press, 2nd edn.

Shankar, R. (1994). *Principles of Quantum Mechanics.* New York: Plenum.

Shankar, R. (2014). *Fundamentals of Physics: Mechanics, Relativity, and Thermodynamics.* New Haven: Yale University Press.

Shima, Hiroyuki and Tsuneyoshi Nakayama (2010). *Higher Mathematics for Physics and Engineering.* Berlin: Springer-Verlag.

Sider, Theodore (2011). *Writing the Book of the World.* Oxford: Oxford University Press.

Sider, Theodore (2020). *The Tools of Metaphysics and the Metaphysics of Science.* Oxford: Oxford University Press.

Singer, Stephanie Frank (2001). *Symmetry in Mechanics: A Gentle, Modern Introduction.* Boston: Burkhäuser.

Sklar, Lawrence (1974). *Space, Time, and Spacetime.* Berkeley: University of California Press.

Sklar, Lawrence (1982). "Saving the Noumena." *Philosophical Topics* 13, 89–110.

Sklar, Lawrence (2013). *Philosophy and the Foundations of Dynamics.* Cambridge: Cambridge University Press.

Slowik, Edward (2005). "Spacetime, Ontology, and Structural Realism." *International Studies in the Philosophy of Science* 19, 147–66.

Slowik, Edward (2016). *The Deep Metaphysics of Space: An Alternative History and Ontology Beyond Substantivalism and Relationism.* Switzerland: Springer.

Smith, George (2008). "Newton's *Philosophiae Naturalis Principia Mathematica*." *Stanford Encyclopedia of Philosophy* (Winter 2008 Edition).

Spivak, Michael (2005). *A Comprehensive Introduction to Differential Geometry*, vol. 1. Houston: Publish or Perish, 3rd edn.

Spivak, Michael (2010). *Physics for Mathematicians: Mechanics I.* United States: Publish or Perish.

Stachel, John (1993). "The Meaning of General Covariance: The Hole Story." In John Earman, Allen I. Janis, Gerald J. Massey, and Nicholas Rescher, eds., *Philosophical Problems of the Internal and External Worlds: Essays on the Philosophy of Adolf Grünbaum*, pp. 129–60. Pittsburgh: University of Pittsburgh Press.

Stein, Howard (1970a). "Newtonian Space-Time." In Robert Palter, ed., *The* Annus Mirabilis *of Sir Isaac Newton, 1666–1966*, pp. 258–84. Cambridge, Mass.: MIT Press.

Stein, Howard (1970b). "On the Notion of Field in Newton, Maxwell, and Beyond." In Roger H. Stuewer, ed., *Historical and Philosophical Perspectives of Science: Minnesota Studies in the Philosophy of Science*, vol. 8, pp. 264–87. Minneapolis: University of Minnesota Press.

Stein, Howard (1977a). "On Space-Time and Ontology: Extract from a Letter to Adolf Grünbaum." In John S. Earman, Clark N. Glymour, and John J. Stachel, eds., *Foundations of Space-Time Theories: Minnesota Studies in the Philosophy of Science*, vol. 8, pp. 374–402. Minneapolis: University of Minnesota Press.

Stein, Howard (1977b). "Some Philosophical Prehistory of General Relativity." In John S. Earman, Clark N. Glymour, and John J. Stachel, eds., *Foundations of Space-Time Theories: Minnesota Studies in the Philosophy of Science*, vol. 8, pp. 3–49. Minneapolis: University of Minnesota Press.

Susskind, Leonard and Art Friedman (2017). *Special Relativity and Classical Field Theory: The Theoretical Minimum*. New York: Basic Books.

Susskind, Leonard and George Hrabovsky (2013). *The Theoretical Minimum: What You Need to Know to Start Doing Physics*. New York: Basic Books.

Swanson, Noel and Hans Halvorson (2012). "On North's 'The Structure of Physics.'" Unpublished manuscript. Available at http://philsci-archive.pitt.edu/9314/.

Symon, Keith R. (1971). *Mechanics*. Reading, Mass.: Addison-Wesley, 3rd edn.

Synge, J. L. and A. Schild (1978). *Tensor Calculus*. New York: Dover.

Szekeres, Peter (2004). *A Course in Modern Mathematical Physics: Groups, Hilbert Space and Differential Geometry*. Cambridge: Cambridge University Press.

Talman, Richard (2000). *Geometric Mechanics*. New York: John Wiley & Sons.

Taylor, John R. (2005). *Classical Mechanics*. Sausalito, California: University Science Books.

Teh, Nicholas J. and Dimitris Tsementzis (2017). "Theoretical Equivalence in Classical Mechanics and its Relationship to Duality." *Studies in History and Philosophy of Modern Physics* 59, 44–54.

Teitel, Trevor (forthcoming). "What Theoretical Equivalence Could Not Be." *Philosophical Studies*. Available at http://philsci-archive.pitt.edu/18875/.

Teller, Paul (1991). "Substance, Relations, and Arguments about the Nature of Space-Time." *Philosophical Review* 100, 363–97.

Temple, G. (2004). *Cartesian Tensors: An Introduction*. Mineola, New York: Dover.

Thorne, Kip S. and Roger D. Blandford (2017). *Modern Classical Physics: Optics, Fluids, Plasmas, Elasticity, Relativity, and Statistical Physics*. Princeton: Princeton University Press.

Torretti, Roberto (1996). *Relativity and Geometry*. New York: Dover, corrected republication of second printing of first edn.

Torretti, Roberto (2019). "Nineteenth Century Geometry." *Stanford Encyclopedia of Philosophy (Fall 2019 Edition)*.

Truesdell, C. (1968). *Essays in the History of Mechanics*. Berlin: Springer-Verlag.

Tumulka, Roderich (2006). "A Relativistic Version of the Ghirardi-Rimini-Weber Model." *Journal of Statistical Physics* 125, 821–40.

van Fraassen, Bas C. (1970). *An Introduction to the Philosophy of Time and Space*. New York: Random House.

van Fraassen, Bas C. (1980). *The Scientific Image*. Oxford: Clarendon Press.

van Fraassen, Bas C. (2014). "One or Two Gentle Remarks about Hans Halvorson's Critique of the Semantic View." *Philosophy of Science* 81, 276–83.

Walker, Jearl, David Halliday, and Robert Resnick (2014). *Fundamentals of Physics*. New Jersey: John Wiley & Sons, 10th edn.

Wallace, David (2012). *The Emergent Multiverse: Quantum Theory According to the Everett Interpretation*. Oxford: Oxford University Press.

Wallace, David (2013). "A Prolegomenon to the Ontology of the Everett Interpretation." In Alyssa Ney and David Z Albert, eds., *The Wave Function: Essays on the Metaphysics of Quantum Mechanics*, pp. 203–22. Oxford: Oxford University Press.

Wallace, David (2019). "Who's Afraid of Coordinate Systems? An Essay on Representation of Spacetime Structure." *Studies in History and Philosophy of Modern Physics* 67, 125–36.

Wallace, David and Christopher G. Timpson (2010). "Quantum Mechanics on Spacetime I: Spacetime State Realism." *British Journal for the Philosophy of Science* 61, 697–727.

Weatherall, James Owen (2016a). "Are Newtonian Gravitation and Geometrized Newtonian Gravitation Theoretically Equivalent?" *Erkenntnis* 81, 1073–91.

Weatherall, James Owen (2016b). "Understanding Gauge." *Philosophy of Science* 83, 1039–49.

Weatherall, James Owen (2017). "Category Theory and the Foundations of Classical Space-Time Theories." In Elaine Landry, ed., *Categories for the Working Philosopher*, pp. 329–48. Oxford: Oxford University Press.

Weatherall, James Owen (2019a). "Theoretical Equivalence in Physics: Part 1." *Philosophy Compass* 14(5).

Weatherall, James Owen (2019b). "Theoretical Equivalence in Physics: Part 2." *Philosophy Compass* 14(5).

Weatherall, James Owen (forthcoming). "Classical Spacetime Structure." In Eleanor Knox and Alastair Wilson, eds., *The Routledge Companion to Philosophy of Physics*. Routledge. Available at http://philsci-archive.pitt.edu/13227/

Weyl, Hermann (1952a). *Space–Time–Matter*. United States: Dover, 4th edn. Translated by Henry L. Brose.

Weyl, Hermann (1952b). *Symmetry*. Princeton: Princeton University Press.

Wilhelm, Isaac (2021). "Comparing the Structures of Mathematical Objects." *Synthese*. https://doi.org/10.1007/s11229-021-03072-0

Wilson, Jessica (2007). "Newtonian Forces." *British Journal for the Philosophy of Science* 58, 173–205.

Wilson, Mark (1993). "There's a Hole and a Bucket, Dear Leibniz." *Midwest Studies in Philosophy* 18, 202–41.

Wilson, Mark (2009). "Determinism and the Mystery of the Missing Physics." *British Journal for the Philosophy of Science* 60, 173–93.

Wilson, Mark (2013). "What is 'Classical Mechanics' Anyway?" In Robert Batterman, ed., *The Oxford Handbook of Philosophy of Physics*, pp. 43–106. Oxford: Oxford University Press.

Wilson, Mark (forthcoming). "Newton in the Pool Hall: Subtleties of the Third Law." In Chris Smeenk and Eric Schliesser, eds., *The Oxford Handbook of Newton*. Oxford University Press.

Worrall, John (1989). "Structural Realism: The Best of Both Worlds?" *Dialectica* 43, 99–124.

Wüthrich, Christian (2019). "The Emergence of Space and Time." In Sophie Gibb, Robin Findlay Hendry, and Tom Lancaster, eds., *The Routledge Handbook of Emergence*, pp. 315–26. New York: Routledge.

Index